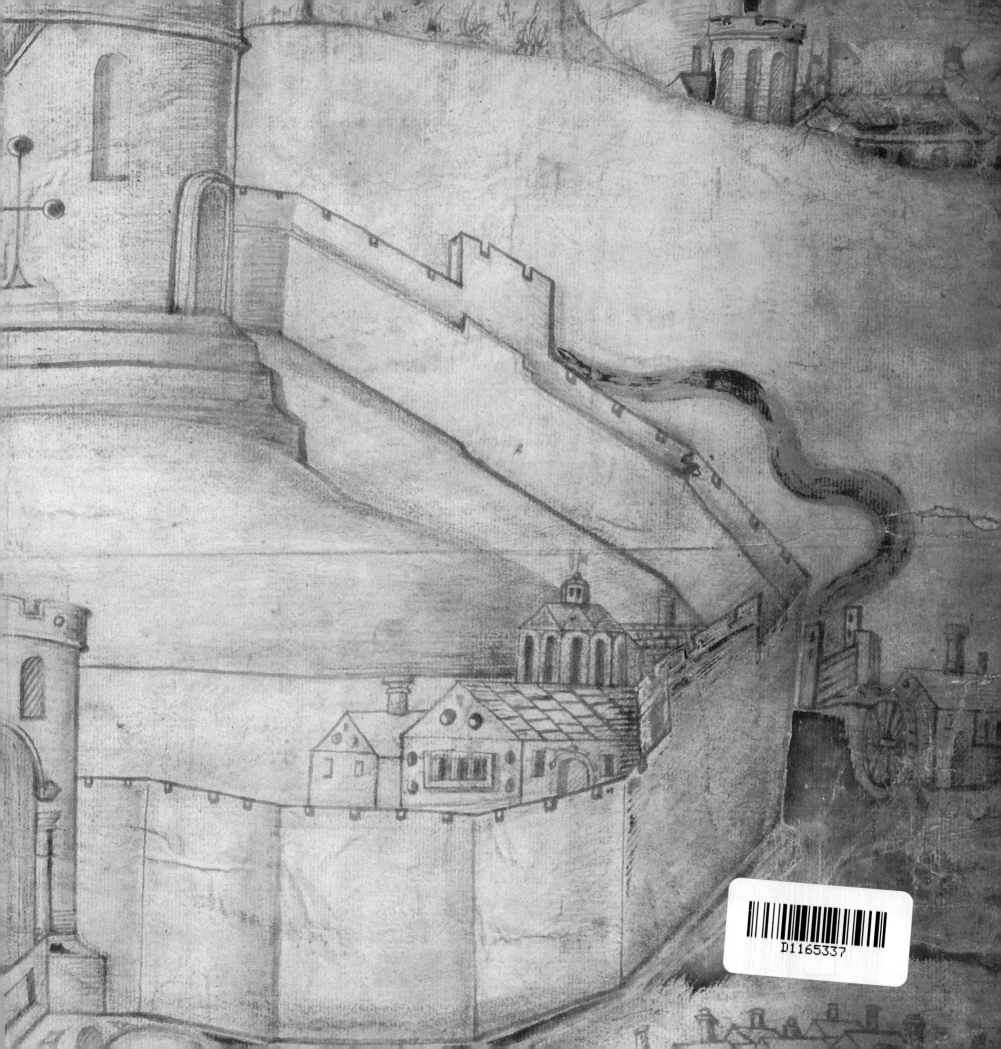

# MAPS
## their untold stories

Rose Mitchell and Andrew Janes

# MAPS
## their untold stories

Map treasures from The National Archives

BLOOMSBURY

LONDON · NEW DELHI · NEW YORK · SYDNEY

The National Archives

Published 2014 by Bloomsbury Publishing Plc,
50 Bedford Square, London WC1B 3DP
www.bloomsbury.com
Bloomsbury is a trademark of Bloomsbury Publishing Plc

ISBN 978-1408-1-8967-2

A CIP catalogue record for this book
is available from the British Library

Design: Nicola Liddiard, Nimbus Design

Printed in Singapore by Tien Wah Press.

10 9 8 7 6 5 4 3 2 1

# MAPS
## out of the archives

Our aim in writing this book was to share more widely some of the remarkable maps here at The National Archives, and also to relate some of the fascinating tales that lie behind them. We have selected 100 maps from among the many possibilities, with a broad range of dates, places and contexts. These maps are not a representative sample but chosen to convey something of the diversity of maps in the archives. We have arranged them into eight themed chapters, although many of our choices could have fitted more than one theme and there are many overlaps in subject matter between maps in different chapters.

We have then placed the maps in their historical and documentary context, as part of the official government archives of the United Kingdom. However, this is to some extent our personal view of them, and we hope that you will perceive some of them differently from us. Above all, we hope that you will enjoy looking at and reading about our maps as much as we have enjoyed writing about them.

Rose Mitchell and Andrew Janes
Map archivists
Kew, London, April 2014

For further information about maps in The National Archives, please visit *www.nationalarchives.gov.uk/maps*

# Contents

# MAPS
## for everyone

There is an old saying that every picture tells a story. We believe that this is even truer of maps than of ordinary pictures. Maps capture and convey knowledge and ideas about places succinctly. It is often easier and more intuitive to make sense of spatial relationships visually from a map than from a description in words alone. Maps also incorporate different layers of meaning. As well as geographical information such as the layout of a city or a piece of land, they often communicate in more subtle ways, such as through their decoration.

Although most people can recognise a map when they see one, scholars of cartography disagree about exactly what the defining characteristics of maps are. It is generally accepted, however, that maps are not completely realistic in the way that photographs are. Instead they represent places selectively and symbolically. They are not always drawn accurately to scale, nor do they invariably portray real places. Not everything that is map-like is necessarily called a map. Maps of smaller areas are often called plans, and maps of the sea

or the air are properly called charts. Whatever we choose to call them, maps are in essence a means of understanding places and making sense of the world.

Old maps are a form of historical evidence that offer a distinctive perspective on the past. They reveal much about the relationships between places and people and they reflect the values, preoccupations and world views of their creators and the context from which they originally emerged. Some maps show how people have left their mark on the environment, by developing land for houses, farming or mining, or simply by travelling from place to place. They may record disputes over land – whether between the inhabitants of neighbouring cottages or the governments of neighbouring countries – and how they were resolved. Other maps reflect the history of how people have explored the world and the discoveries that they have made. They portray how places and landscapes have been owned, settled, exploited, attacked and defended, and how empires rose and fell. We can also enjoy and appreciate maps for their own

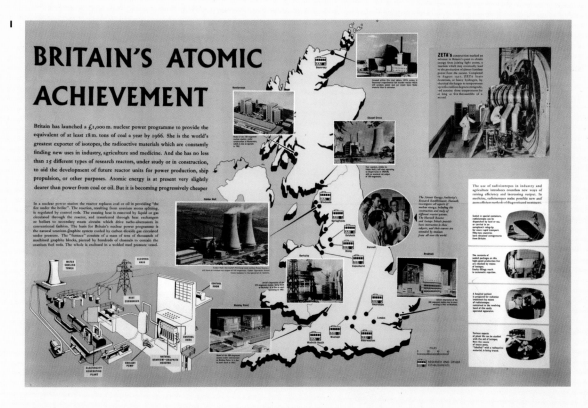

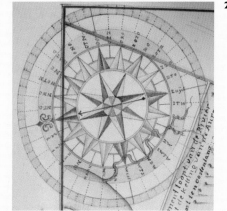

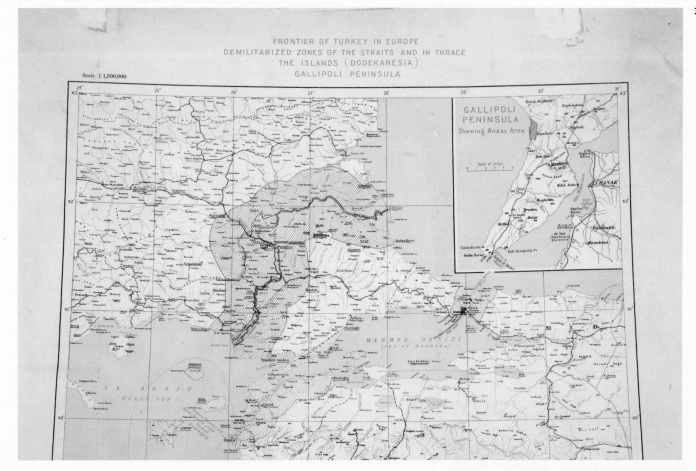

**1** NUCLEAR NATION: This 1950s poster expressing pride in British scientific achievement includes a map as part of its overall design.

**2** WHICH WAY IS UP? Compass indicators like this one are decorative features with a practical purpose.

**3** MAPPING THE BOUNDARIES: Simple lines on a map can matter a great deal to the people affected by their equivalent on the ground. This example shows the new boundaries of Turkey, agreed under the Treaty of Lausanne in 1923.

sake, because they are intrinsically interesting in themselves.

Our map collection at The National Archives is one of the finest and most remarkable in the world. It is also one of the largest, containing more than six million maps and similar items, such as topographical views and architectural drawings. To describe it as a 'collection' is actually something of a misnomer, because our maps are treated not as a separate entity but as an integral part of the archives. They often accompany and illustrate documents such as letters or reports, to which they may be intellectually and sometimes physically attached. Information from these related records helps us to interpret the maps and to understand the stories that lie behind them.

Just like the other historical records in the archives, the maps now in our care were created or used by the government of the United Kingdom. They illustrate the great themes of history from the British perspective, as well as day-to-day administration at home and overseas. The government has needed maps most commonly in relation to warfare and defence, international relations, diplomacy, and the impact of the state on the physical environment, but it has used maps for many other purposes too. From the assertion of land ownership to the determination of boundaries, and from tax assessment to the administration of justice, maps have made history by influencing government decisions, both great and small.

Our maps show every corner of the British Isles and places all around the globe. Their geographical spread reflects the British government's strategic, political and administrative interests and priorities. For instance, we have many more maps of Guyana, which was formerly a British colony, than we do of Paraguay, which has few historical ties to Britain. Even within the United Kingdom, certain places – such as Crown Estate land and important naval bases – are represented more frequently than others. Coverage is equally uneven in terms of date. Although the oldest of our maps were drawn during the 14th century, the vast majority date from the

**4** PICTURE THIS: Some maps are very abstract and others more pictorial. This view of Walland Marsh in Sussex, probably drawn in 1536, has as much in common with a painting as it does with a modern map.

**5** PAPER LANDSCAPE: Fort Albert and its surroundings on the tiny Channel Island of Alderney come vividly to life in this hand-drawn map dating from 1869.

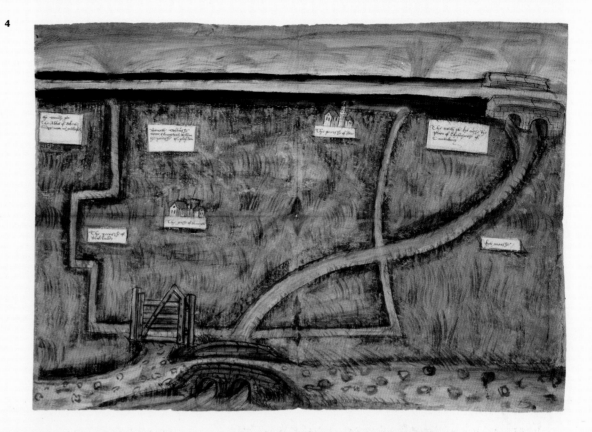

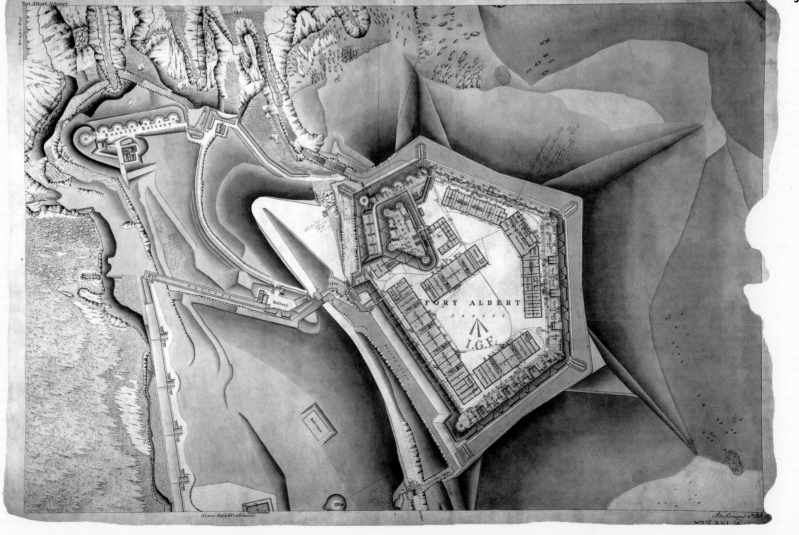

18th, 19th and 20th centuries. Many areas of the world – including some parts of England – were not mapped accurately in detail until the 19th century, or in some instances even more recently.

The mapmakers whose work is found in the archives were exceptionally diverse. Many were professional surveyors, cartographers or printers, some employed by the government itself (either as civilians or in the armed services) and others working for commercial enterprises. Maps were also drawn by amateurs, such as British officials or travellers overseas. Some of the maps included were created by people from other countries who came into contact with British colonists, diplomats or explorers.

The maps themselves are no less varied than their makers and the places that they portray. Some are colourful and imaginative, others quite plain. Very many are unique, either because they are original manuscript creations or because they are printed items annotated by hand. Manuscript and printed maps co-exist among our records from the 16th century onwards. Even in the 20th century, maps were still sometimes entirely hand-drawn where few or no additional copies would be required. It became increasingly common, however, for the government to amend or customise an existing published map – often one produced by its own Ordnance Survey or armed forces – in preference to creating an entirely new one.

Although the archives were accumulated by the government of the United Kingdom, they include far more than just the stories of the powerful and the governing elite. Our records encompass a broad variety of human experience, from great events to the challenges and achievements of everyday life. They also contain the histories of longer-term forces, such as social and political change and the evolution of the landscape. These are the stories of people and places throughout Britain and the world, and they are also the stories of our maps.

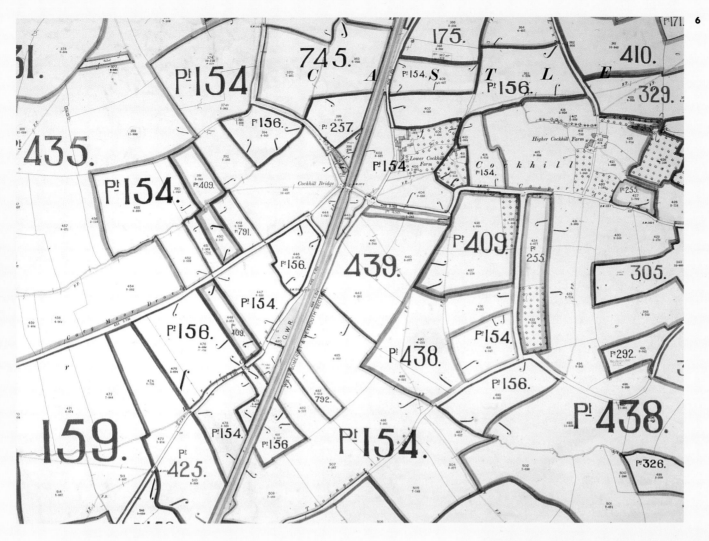

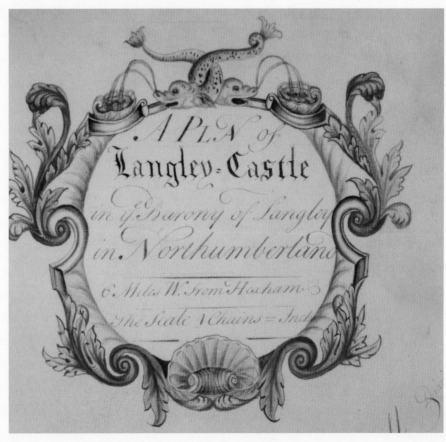

**6** UNIQUE ADDITIONS: This Ordnance Survey map was annotated as part of the Valuation Office survey, carried out in England and Wales shortly before the First World War. These records are particularly popular with genealogists and local history researchers.

**7** DECORATIVE TOUCH: Fish and seashells adorn this cartouche from an 18th century estate map. Many landowners valued maps as much for their beauty and prestige as for their utility.

# CHAPTER ONE

# EARLY maps

Maps from earlier times are different from the kind of maps we are familiar with today. In this chapter we explore a selection that illustrates the diversity of maps from the 15th century to the 17th century, and some of the reasons why they were made. Most of these maps were manuscript, drawn before more standardised printed maps became common. They are often colourful and provide highly individual views of places, buildings, plant life and coastlines, before the genre of British landscape painting had begun. Sometimes mapmakers tried to show hypothetical places such as a posited southern continent of the same size as the Arctic, and the map on page 43 even shows the outer edges of the heavens.

Maps gradually became accepted as a useful and attractive way to present viewpoints, knowledge and information. The oldest maps in the archives date from the 14th century, but maps were relatively rare until the 16th century. The archives holds seven maps and one chart dated before 1500. Two of these maps are included in this chapter, showing medieval landscapes in Surrey and Yorkshire. The number of maps produced increases to about 75 for 1500–1550, doubled for 1550–1580, and with perhaps 200 further maps for 1580–1600 when mapmaking became more widespread.

Why did people start making and using maps? From the early 16th century maps began to be increasingly appreciated and used for diplomacy, defence and government administration. Map use in law courts was encouraged after a map affected the verdict in a case in 1515. Influential individuals such as the statesman Lord Burghley saw the importance of maps and helped to make them common currency at the court of Elizabeth I. The visual nature of maps was especially helpful in conveying information about far-off places, when planning military campaigns abroad, colonising, and for an overview of foreign states, especially those which were likely to be a threat to England. See, for instance, the French administrative map on page 45, used for military intelligence.

Early maps were usually made for a specific

**1** WOLVES, RABBITS AND WILD MEN:
A scene in the hills above Portrush peninsula,
County Antrim, Ireland in 1580.

**2** CASTLE IN THE SAND: Camber Castle
near Rye, Sussex (see page 39).

**3** LAKELAND FELLS: Dynamic depiction of
hills surrounding the river valley and hamlet
of Sadgill, Westmorland, drawn for a court
case in 1578.

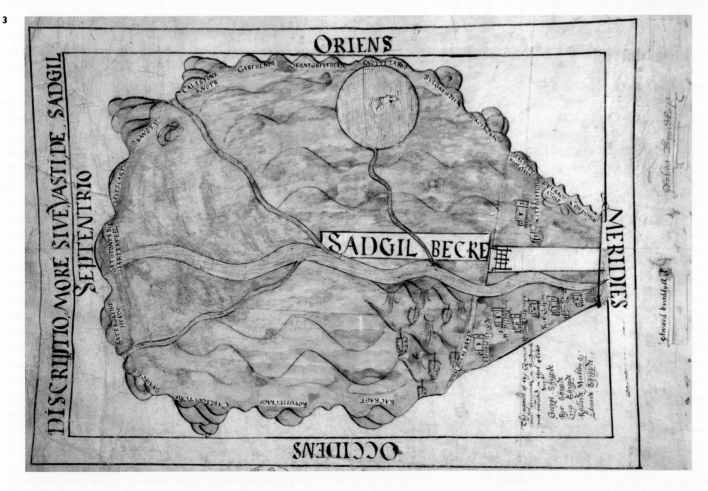

purpose, by hand, when needed. They dealt with a particular issue where multiple copies were not required, or with secret matters such as defence where they were for the eyes of the ruling elite alone. What is shown is selective, chosen to reflect a map's purpose. The Knaresborough Castle plan (page 33), drawn to show its defensive state, reveals little of its surroundings, while we might wish that the map of Cheam's common fields (page 31) could show more of the nearby palace. Maps drawn for opposing parties in a legal case may show very different views of the same land. This limited focus on the matter in hand resulted in maps of small places such as remote moorlands but fewer of towns, regions or the whole country.

The success of early maps was measured by the extent to which they achieved their purpose, rather than their accuracy in the modern sense. They may not be drawn to scale, and the mapmaker might even emphasise important features by portraying them disproportionately larger than their true size. Some maps are rough sketches, others in pictorial style. They might be drawn to make best use of the piece of parchment or paper and so north may not be at the top. We may need to look at related documents to interpret a map, to tell us who made it, when, why, and what happened to it next, questions which early maps often do not answer in themselves.

If reading early maps requires skill and a fresh mind, it also offers us a window on the world of centuries past. It can be fascinating to glimpse lost palaces and castles, long-fallen ancient crosses and deserted villages, in a landscape often vastly changed from the one we see now. Until the last quarter of the 16th century mapmaking was an

**4** A TRUE AND PERFECTE PLATTE:
Manuscript maps were usually called plattes
during the 16th century.

**5** GALLEONS IN ACTION: These ships in
full sail with cannons ablaze are drawn in the
sea on a map of Ireland (see page 35).

**4**

**5**

activity to which any educated man might turn his
hand, with no particular training in cartography but
a knowledge of the place that he drew. Mapmakers
used their creativity to express the point of the
map, and added embellishments in the form of
animals, monsters and ships. Their maps were made
in a profusion of styles, shapes and colours.

The rise of a class of professional surveyors
using new techniques and survey instruments
brought a new era of mathematically constructed
work. These men were working to commission and
needed to please their employers and advertise
their skills. They included scale bars and compass
indicators, and titled their creations to explain what
they showed and why they were made. Although
these new cartographic skills were not employed
consistently, this period transformed the art of
making maps towards the kind we know today.

# View from a medieval monastery

CHERTSEY ABBEY, SURREY c.1430

Why make a map? In the Middle Ages, reading and writing were rare skills and largely the preserve of monks, who carefully kept copies of important documents such as charters and deeds about their monastery's estate bound together in a volume called a cartulary. Maps such as this one began to be made, as a different and effective way of showing ownership and rights over land.

After the Benedictine abbey at Chertsey was involved in a property dispute in the 1420s, the monks left an account for future custodians of the Abbey about its claims and where the lands in question lay. This map was drawn in the cartulary to illustrate points made in their text, and readily conveys important features in the case. The wisdom of the monks in creating this record against future need was borne out when the cartulary was called as evidence in later times; a note on the flyleaf states that it was deposited in the Exchequer Court in 1637.

The dispute was about grazing rights on the Abbey's pastures and meadows at certain times of the year, claimed by some of the tenants for longer than the Abbey allowed. The map notes the size and names of these low-lying fields, prone to flooding, and forming an island enclosed by the River Thames. A smaller waterway cut across the land, dug by the monks to drive the Abbey's water mills. The fields are drawn fairly conventionally in plan, and the other features of the map are in their appropriate position in relation to each other, although not to a consistent scale.

The map has some of the quality of a picture, especially in the way it shows buildings: houses in the village of Laleham across the river, the large barn above the church in which grain from the fields would have been stored, and two mills to its right, either side of the river, in which grain was ground. The far mill's wheel is drawn facing the viewer, while just the top of the wheel of the near mill is visible. The Abbey's cluster of buildings is shown by a detailed elevation of its church, with lit interior and open door, and drawn disproportionately large, as if to emphasise the map's provenance; as the church itself must have towered over the flat landscape.

The colours on this map have kept their brightness across more than half a millennium, through being kept in the dark, enclosed in the cartulary. The red tile roofs sing out against the grey lead of church roof, the blue waterways, green vegetation, and the wood of Chertsey Bridge at right. It must have seemed to the monastic mapmaker that the landscape he drew, dominated by the church materially and socially, was set to last forever. Yet the church he knew would be swept away in the next century by Henry VIII's Dissolution of the Monasteries, leaving this map, as with so many others in the archives, as a record of landscapes and times past.

SEAL OF CHERTSEY ABBEY: This 11th century depiction of the Abbey church on its seal shows a building which differs in detail and angle from that on the map.

DOMESDAY RECORD: The entry for Chertsey Abbey in the Domesday Book of 1086 shows that it had extensive estates.

# Bells, bridges and bog plants

Which way up is this 15th-century map meant to be seen? North was not to become fixed at the top of a map for several centuries, and while the direction of the cardinal points are given, the words written on the map face different ways. Perhaps it was intended to be seen from multiple viewpoints, by parties in a legal case seated around a table. It was made to record a dispute over villagers' rights to cut peat for fuel and to pasture their animals on the central area called Inclesmoor (written in red 'upside down' across the moor), some of which remains as Thorne Moor and Goole Moor just east of Scunthorpe, on the Yorkshire-Lincolnshire border.

The map is drawn on parchment, with the distinctive shape of the animal's neck to the right. It covers an area of about 240 square miles. Inclesmoor is shown surrounded by villages and enclosed by a network of rivers: the Don, Ouse, and Trent join the Humber in the north-east corner (lower left), while the Aire flows along the western edge of the map (on the right), crossed by a road carried on a stout stone bridge. These rivers are imaginatively drawn so that they seem to roll up like carpets at the map's edge, bounding the moor in the middle so that it resembles an island.

Within this striking design the map beguiles with detail of both the built and natural environment. The central moor and surrounding pastures are drawn like a tapestry, covered in marshland plants, while trees crowd round the villages, and rather impressionistic willows appear as smudges along the rivers. Churches and houses, bridges in solid stone or mere planks, and wayside crosses, are all drawn in bird's-eye view, so that the viewer appears to be looking down on them. Each village is shown with realistic buildings rather than symbolic images. Timber-framed houses have different cruck patterns, roofs may be thatched or tiled, and churches are shown with spires or towers. Some features appear larger than others; a stone cross just to the left below the moor is drawn larger than the village next to it, presumably because it was an important locus in the dispute.

Another version of this map, smaller and simpler in design, was drawn in an early 15th century volume with other documents kept as a record of the dispute in case it arose again. The Duchy of Lancaster was one of the parties to the dispute, with St Mary's Abbey at York, and both maps passed to the Duchy's archives, now held at The National Archives. Whether the smaller map was a distillation of the essential features of this larger, and therefore earlier or contemporary one, or the larger map was a later elaboration of the smaller one, the mapmaker knew the terrain that he drew in such detail from life, presenting through one of relatively few surviving maps from this time a striking vision of the medieval rural landscape.

**NOW-DESERTED VILLAGE:** Haldenby village, by the River Don, was deserted later in the 15th century, but here it is shown with several houses among trees, and the church is one of three on the map where a bell was drawn hanging in the tower.

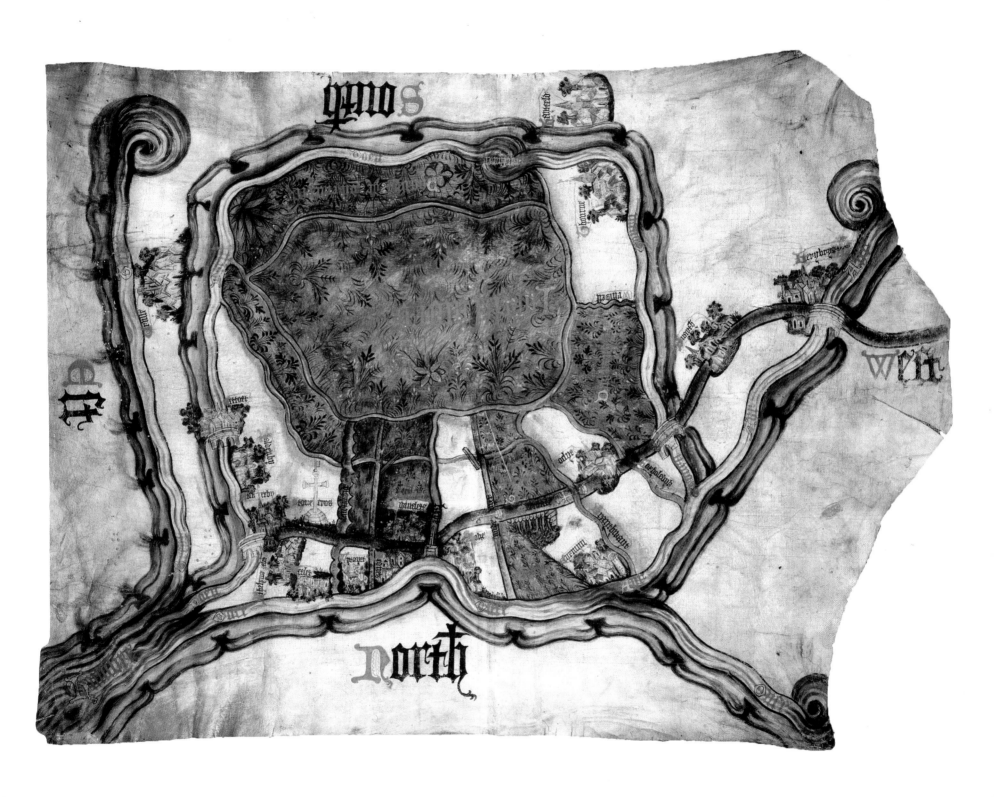

# A deserted hamlet

NEAR TUNBRIDGE WELLS, KENT, c.1519

It is wise not to judge a map by its looks. Pen and ink sketch maps like this one are just as likely to convey useful or interesting information as more attractive creations. Drawn for a dispute over the lands shown, this small map (less than 40cm by 60cm) incorporates a wealth of detail, including a county boundary, an otherwise unrecorded deserted hamlet, the text of a fourteenth-century document about the land, a royal hunting forest of Henry VIII and dower lands of one of his queens.

Although few of the places named on the map still exist, the area shown was in the vicinity of what is now Tunbridge Wells, long before the discovery of mineral waters led to the growth of the spa town in Georgian times. At the top right of the map runs the boundary between Kent and Sussex. At the lower edge, which is marked 'north', a wavy line indicates the park pale of Southfrith, a royal deer park. The focus of the map is a group of buildings which bear a place-name no longer known: Bromelerge. The large house is labelled as that of 'Wybarn': Anthony Wybarn, who claimed some of the lands shown on the map. Below this, four houses are shown with the legend 'All this sometyme was a hamlet called bromelerge'. This may well be the first deserted hamlet to be depicted and described as such on a map.

Other features on the map give an idea of the geographical layout of the area, with roads, a gate through the park pale, and a note to the right of the hamlet, written against a diagonal line, that this hedge was planted within living memory; 'ther be men a lyve that know when this hedge was made'. Field boundaries and acreages were shown, since the case rested upon identification of fields across several centuries. Text on the map gives wording from a grant in 1330 of certain lands in 'Southfrethe' by Lady Clare, who founded Clare College, Cambridge; a copy of this is in the archives. The lands passed to the Crown and, by the time that this map was made, were part of the estate of the queen, Catherine of Aragon. The way that the text is used on the map counters Wybarn's claim, on the grounds that the acreage of claimed fields was wrong. The outcome of the case is not clear; Wybarn's will of 1528 mentions 'londes in Tonbrigge', but does not specify exactly which lands.

There are two even sketchier maps of the same area, apparently drawn by the same hand, but from different angles and showing confusion about their orientation. All this suggests that the person who made this map was not used to doing so. Even if the result here appears rather diagrammatic to our eyes, it does use the visual aspect of mapping to help to clarify where the important features in the case lay. It thus represents a metamorphosis in the recording of land, from written records to maps.

WHERE IS NORTH? An even sketchier map of the same area, apparently drawn by the same hand, has east at the lower edge, instead of north, as in the more detailed version.

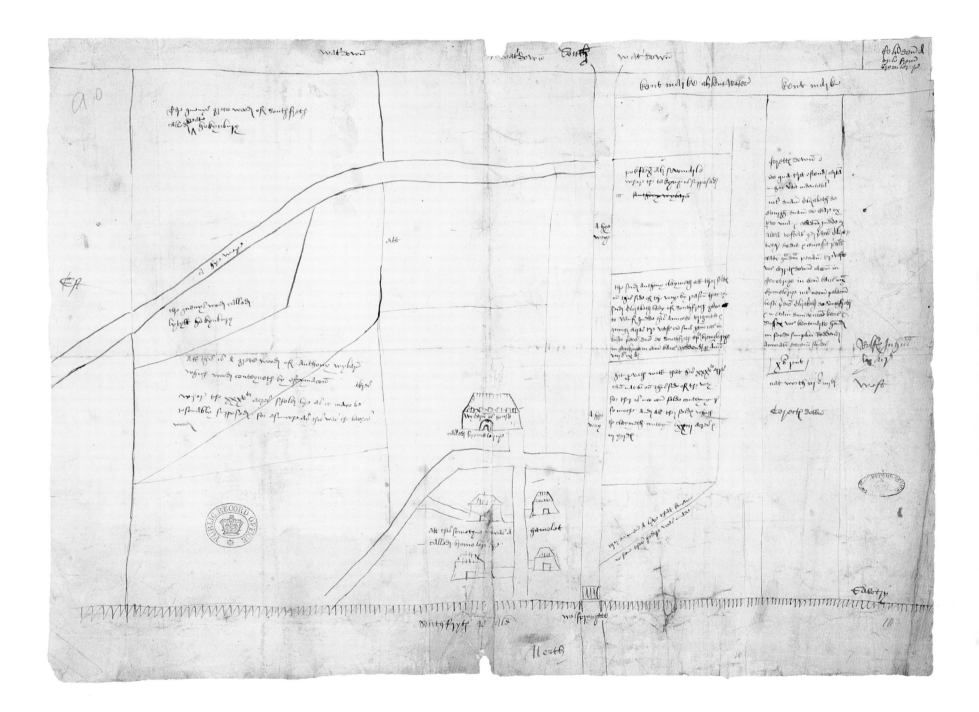

# Castles and commons in a Welsh valley

This map shows an area of just a few square miles in the Vale of Glamorgan, on the south coast of Wales; the sea runs along the left edge of the map. When and why was it made? It can be difficult to date early manuscript maps precisely if there are no papers to which they are obviously related. One method is to use datable evidence, where something on the map can be linked to a specific event. This map pictures Ewenny Priory at lower right (to the north, spelt 'Wenye') looking solid and intact. It was dissolved in early 1540 as part of the Reformation, and much of the building was dismantled. There is also a rare view of Dunraven Castle to the south (top left) which was demolished and replaced by a manor house at some point in the 1540s. These features point to a pre-Reformation date for the map.

The map shows much else besides Ewenny Priory and Dunraven Castle. Commons and downs are coloured brown, the king's wood and another wood shaded green, and roads are drawn in white, with notes about where they led. There is a depiction of the imposing multi-towered Norman castle at Ogmore, above centre right, commanding the confluence of two rivers. Other buildings such as churches and houses are also drawn in perspective, in the villages of St Brides Major and Wick, and in nearby hamlets. However the Priory is shown disproportionately large compared to the other buildings, which suggests that it was significant for the purpose of the map.

There is a second map showing a similar area to this one, but oriented differently, without reference to Ewenny Priory, and with different details, including a prominent mill not shown on this map. We know why and when that one was made. It dates to June 1579, from papers in a case heard in the Court of the Duchy of Lancaster (the Duchy was lord of the manor of Ogmore). Commissioners in that case stated that they made that map, and notes on it explain that it shows areas of common used by householders of St Brides, in which the householders of Wick claimed common rights.

Where there are two apparently similar maps, as here, it is tempting to assume a relationship between them. The use of place names such as 'The kings wood' and 'The kings mill' on the second map, dated well into Queen Elizabeth I's reign, might suggest that the later map is a copy from one of the reign of Henry VIII or Edward VI. Did the commissioners perhaps use an earlier or extant map – this one - as a base? If so, it seems strange that they do not mention another map besides their own. If these were a draft and a finished map, they would be likely to resemble each other more closely. Perhaps more evidence may yet appear to throw new light on this map and help to solve its dating puzzle.

DISSOLVED PRIORY: Ewenny Priory, shown here intact, was dissolved on 6 January 1540, and much of the structure demolished.

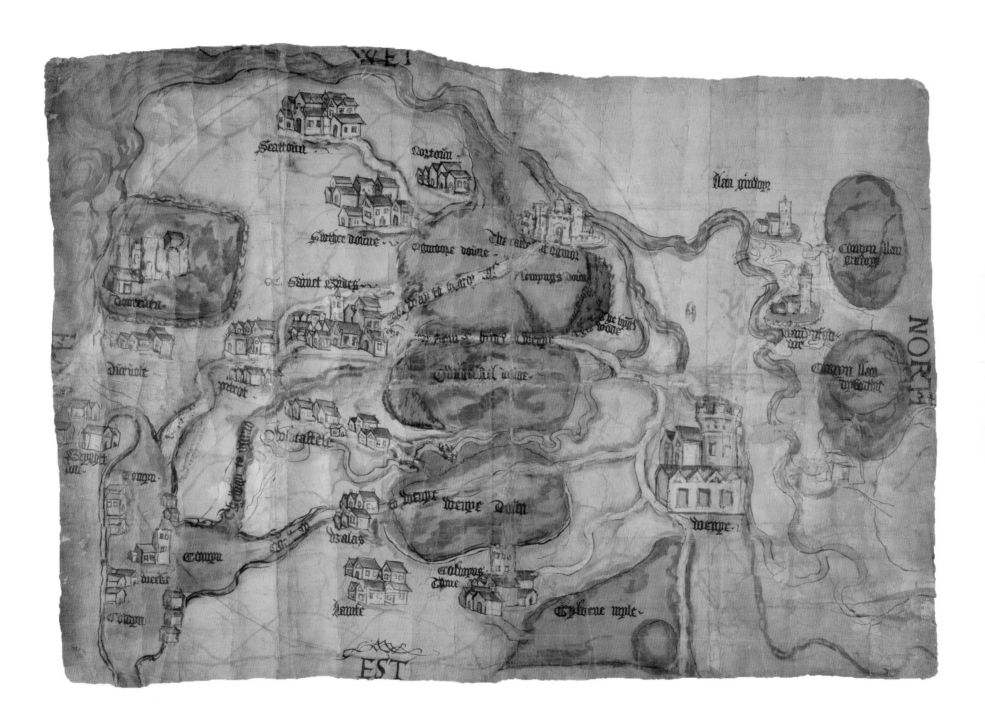

# A border line case

When this map was drawn in 1552, England and Scotland were separate countries, their crowns not united until King James of Scotland succeeded Elizabeth I in 1603. These two nations were not always peaceable neighbours, so the Anglo-Scottish border needed to be fixed, especially at the western end, which was a stronghold of lawless clans. This was the 'Debatable Land' between Cumberland and Dumfriesshire, shown in green on this map.

The area was nearly four miles wide, lying between the rivers Sark and Esk, their waters running into the Solway Firth at left. Gretna is off the map to the left, while Canonbie is at centre. The map shows a wild land of hills, moors, bogs or mosses, and smaller burns running off rivers. Note the way that shading is applied outside the hills, as if they have hair. A standing stone near the word 'Northe' at top right is about ten miles from the southern Lochmaben stone on the Esk estuary. The latter was the traditional 'trysting place' where the warring factors met to negotiate truces and exchange prisoners. The fortified towers of two such prominent families are shown on this map: those of Fergus, Tom and Richard 'Greme' (the English Grahams) at centre, and the Scot Sandy Armstrong's tower between the left two lines at the top.

This important strategic map was made by Henry Bullock, otherwise recorded only in connection with buildings, as he later became Master Mason of the King's Works. Chosen as 'a man of some experience', he was paid 20 nobles (about £1,500 today) 'for special service' – presumably for 'the juste and true making' of this map. Perhaps as a response to the demands of high-level politics, this is an early example of a map drawn to scale, with a scale bar at its lower edge. It appears that Bullock used a local or 'customary' scale; a perch of about 8 yards long instead of the standard 5½ yards can be calculated from the length of the four straight lines on the map against their length on the ground.

These lines illustrate English and Scottish proposals for the border, a compromise suggested by the French, and 'the last and Fynal Lyne' agreed on 24 September 1552. The map was filed with an account of the negotiations by the French ambassador in London, a mediator in the case. While the final line gave the largest area of the Debatable Land to Scotland, this was mostly moorland. The new frontier was then marked on the ground by an earthwork from Esk to Sark, still evident as Scots' Dike.

State Papers reveal that drawing the line on this map was considered to be an integral part of defining the border. It is an early record of a map used as a tool in making an international boundary, which gives us a visual idea of the process behind it. The lines represent the debate about the land among the three parties.

DRAWING THE LINE: Any one of these four lines might have become the Anglo-Scottish boundary. The top line was the English proposal, with the Scottish 'offerr' parallel; the French suggestion ran diagonally between them. The final line agreed has at each end a cross pattée with wedge-shaped arms.

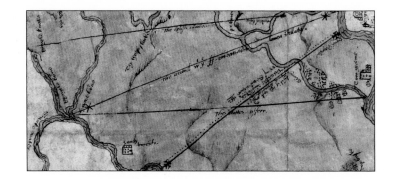

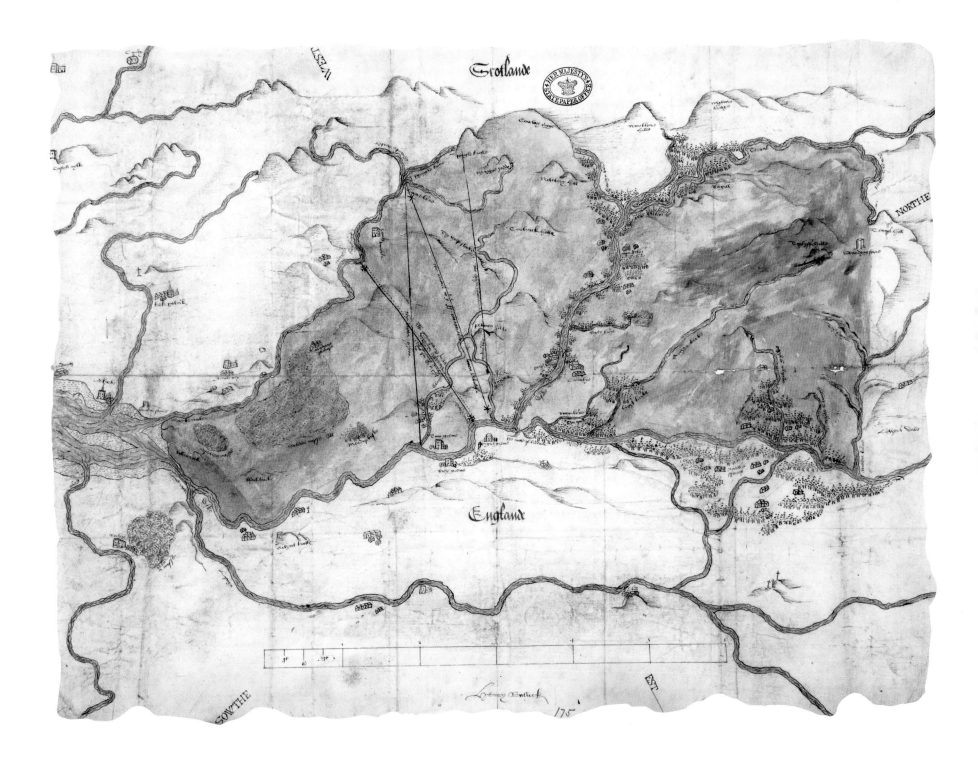

# In the shadow of a long-lost palace

CHEAM, MALDEN, MORDEN AND NONSUCH; SURREY, 1553

This map only incidentally shows Henry VIII's spectacular palace of Nonsuch, in the top left border, with the paling and gates to its parks across the top of the map, and what may have been a prospect mound from which royal parties could view the hunt. Nonsuch was a favourite retreat of Elizabeth I, but was demolished by Charles II's mistress who sold the materials to pay her gambling debts. The park remains as a remnant of past glory, near Epsom in the commuter belt south-west of London. The palace was still new and not quite finished when this map was made, and must have had enormous impact on its surrounds both physically and symbolically. Indeed, it completely erased a village on its chosen site.

Yet the landscape pictured here just outside its gates suggests an unchanged countryside where life continued in a traditional way. Villages are represented by their church and perhaps a few houses. Fields (coloured brown) are shown in detail, with hedge or hurdle boundaries, and their names or the villages to which they belonged. Individual trees are drawn at field edges or in groups. Spark's Elm, a single tree just above the map's centre, stands as a landmark on the common called Sparrow Field (coloured green, along with paths). This common was both a scene of much activity and the place where village boundaries met, which caused friction over who could do what, and when.

This map was one of two produced in a case in the Exchequer Court of Augmentations, concerning villagers' rights to graze their beasts on the common. The first and plainer map (also in the archives) was submitted to the court by the plaintiff, but deemed not to show the relevant features correctly when taken to the field by the commissioners appointed to investigate the complaint. Maps submitted by parties in a dispute were thus not accepted uncritically, but judged by the criteria of the time. The 'imperfection' of the first map was that it was not sufficiently accurate to convey the facts of the case to a court in London which was distant from, and unfamiliar with, the ground in question.

The commissioners, who were local gentry, then made this second map, which they described as 'a true and perfecte new plot', for 'the more playne manifest and direct understondyng' of the problem. Important and relevant additional details include the drove ways by which villagers took their animals to the common; places where beasts of one village were impounded for trespass by another; and processional paths to crosses on Sparrow Field, used by the villagers of Morden and Cheam when beating the parish bounds.

This map was seen as 'true' in the sense that it was drawn to make the matter clear, although it is not necessarily true in the modern sense of being drawn to scale. It may appear rather quaint to us now, but this map gives an insight into the workings of village life in the Tudor landscape and was thought 'perfect' at the time it was made.

OVER THE PALACE PALE: This may be the earliest known view of Nonsuch Palace, which Henry VIII started building in 1538 with the aim that there would be 'none such' to match it. It shows the turreted gatehouse and buildings inside the outer wall, as seen by bird's eye, from the direction of the avenue between the two parks, which led to London. It is worth looking at surviving maps of the area in which important buildings were located, in case they were shown as landmarks.

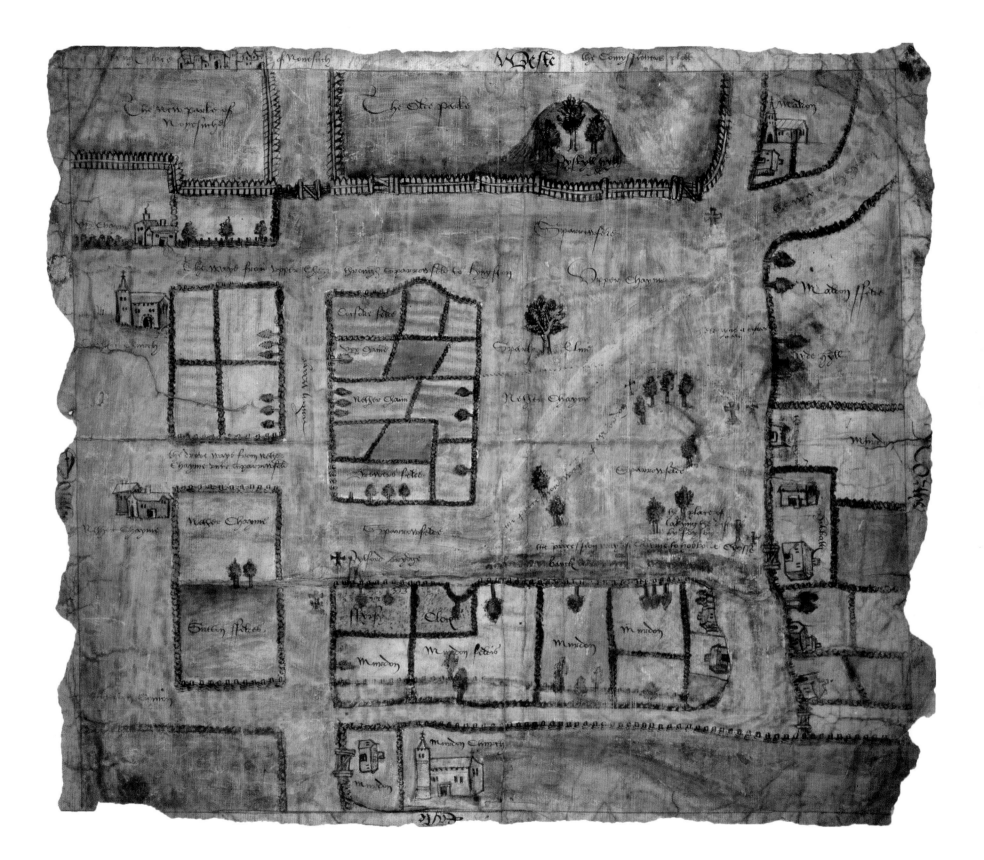

# Beast on the battlements

Defence was a major preoccupation for England through much of the 16th century, with the threat of invasion from France and Spain, and also from Scotland and the troubled north of England. Elizabeth I inherited the problem of unrest in the north, heightened by the return to Scotland from France of her rival Mary Queen of Scots in 1561. In that year Elizabeth ordered a survey of northern castles which belonged to her as part of her Duchy of Lancaster, to find out their soundness against potential attack from Scotland and how much any necessary repairs would cost.

So one summer over 450 years ago the Chancellor of the Duchy of Lancaster, Sir Ambrose Cave, set out from London to spend over two months touring castles, mainly in Yorkshire, with some in Derbyshire and Lancashire. The results were a detailed written report on each castle's state and a portfolio of drawings. It is not known who drew them. Cave did not say that he made them himself, nor did he claim expenses for a mapmaker, but it is possible that an artist accompanied him on his tour. Written surveys were quite common, but it was unusual for maps to be made. Perhaps Cave was influenced in having these drawings created by his kinsman and friend William Cecil, the great Elizabethan statesman and close adviser to the Queen. Cecil was an enthusiastic advocate of maps as useful tools of government, and he made maps himself and annotated many others.

This is one of eight castle drawings extant, and shows Knaresborough Castle in Yorkshire and the area around it. The castle is drawn in elevation as if approached from below the impressive gatehouse, with its portcullis raised. The curtain walls go off into the distance, not drawn in correct perspective, with the height of the far curtain wall exaggerated so it appears more imposing and suggests a large space. The roofs of buildings inside the castle walls are visible, indicating the living quarters. Outside the dry ditch are drawn a watermill, sluice, river, bridge and trees. These represent the Forest of Knaresborough, which stretched around the castle for 30 miles and afforded such good hunting that it had been a favourite of King John in the 13th century.

Cave returned to London and laid his report and drawings before a number of the Queen's Privy Councillors. They used this information to inform their decision to maintain or dismantle the castles, taking into account factors such as the cost of repairs and the likelihood of rebellion in their region. Most were kept, where they provided an occasional royal residence or a gaol, were strongpoints against rebellion, or administrative centres for Duchy estates. Knaresborough was to be 'contynued and kept' for the value of its forests, parks and chases, which supplied deer for the Queen to give her servants in lieu of money or more costly presents. Thus this apparently simple, even naïve, drawing, is evidence for the growing influence of visual material in matters of state.

CASTLE WITH A VIEW: On the roofline to right the artist shows two trumpeters flanking two regal figures with crowns and sceptres; perhaps symbolic of Edward III and his wife Philippa, who spent much time at the castle in the 14th century. The lion was presumably a statue.

# Sea monsters, galleons and misty mountains

This map shows the island of Ireland, as viewed by statesmen in London. It appears rather squashed in shape, until we realise that it has west not north at the top; turn the map to the left to see a more familiar outline. Among the earliest cartographic representations of Ireland, this map was drawn in 1567, at a time when the Protestant Elizabeth I wanted tighter control of Ireland to prevent it becoming a prize for her Catholic enemies on the continent. She colonised by supplanting the native Irish baronies with 'plantations' given to Englishmen. The very act of mapping was part of the appropriation of power over land.

Ireland was up to that time uncharted territory from the viewpoint of its remote English rulers. However, this overview map gives the impression of a land which could be named, drawn, known – and therefore colonised. The island is set within a frame, with the west coasts of England and Scotland at the lower edge, bearing their respective flags. The title, formally given in Latin for an educated readership, can be translated as 'Hibernia, an island not far from England, in the common tongue called Ireland'. It is set in a fretwork cartouche, flanked by the cross of St George, and by a crowned harp denoting the kingdom of Ireland. All of this served to locate Ireland in the world of knowledge familiar to English government.

Drawn just off the Welsh coast, a scale bar surmounted by dividers indicates that the map was drawn at a scale of about 1 inch to 16 miles. Scale was a relatively new concept in mapping at that time, and its inclusion on this map together with degrees of latitude and longitude around the outer edge shows that the mapmaker wanted to give an impression of being familiar with the latest developments. This man was John Goghe, who signed his name in the lower right hand corner, but about whom nothing else is known.

Goghe's map includes details of interest to the English: the location of castles and forts, names of settlements, difficult terrain for armies such as mountains and lakes, the holdings of Irish earls and strategic islands and river crossings. Sea was the main means of transport at that time, and there is fine drawing of the coast, harbours and rivers. The sea is decorated with ships in full sail, some with guns ablaze, while huge fish and sea monsters suggest a sense of danger.

There is evidence that the map was used not just to inform those in power, but also as a working document to help form defence strategy and influence decisions at the highest levels of government. Inked additions in a sketchier hand containing information about extra places and persons of note were made by Sir William Cecil, who became Lord Burghley, one of Queen Elizabeth I's most influential ministers. Maps such as this one were one of the tools of defence and colonisation, along with the written survey and the gun.

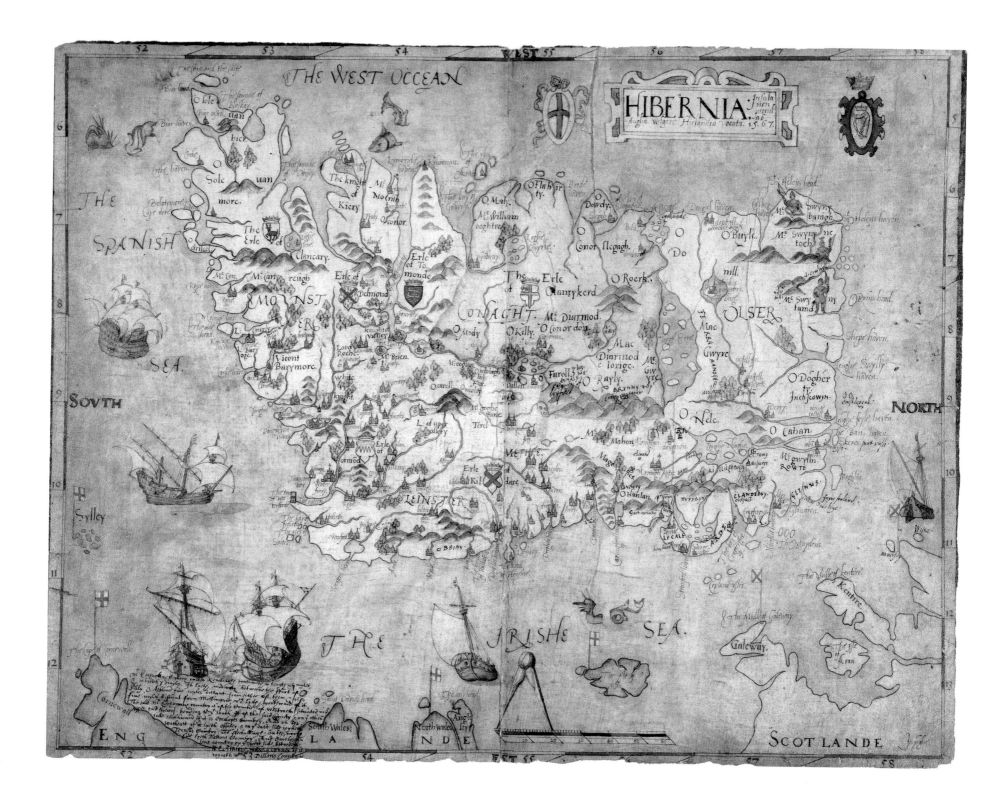

# Summer on the lake   STURMER MERE, ESSEX, 1571

You will not be surprised to learn that this map illustrates a dispute over fishing rights on this reed-fringed lake by the River Stour, on the border between Essex and Suffolk. The water is the focus, with streams and channels leading into the central mere, and several plank bridges to cross from one marshy ground to another. Towards the edges of the map lie the red-roofed manor house at Wixoe (to the south but at top of this map), and houses to the left, towards Stoke-by-Clare. Nearby is a boathouse which had belonged to an ecclesiastical body called Stoke College.

This vivid portrayal of a wetland landscape has added value for ecologists and local historians, as it is covered with notes. These record the state of the ground around the mere, who owned it, what grew on it, and whether it was good land or marsh. Bullrushes are differentiated from 'flag reeds', and woodland from meadow. At top left, land called 'the Wett Fenn' is noted as an alder wood. All this detail was needed, as the boundaries of the mere were hard to define, and fluctuated with the seasons and weather. The map must have been made in summer, as a note tells us that what is shown as ground below the mere was so flooded in winter that a number of boats could then row upon it.

The plaintiff in the lawsuit, Thomas Barnardiston, claimed that he had inherited fishing rights both in the river which flowed into the mere, and 18 feet into the mere. However, Stoke College had previously claimed all fishing rights over the mere and, when the College ceased to exist in 1548 on the dissolution of the monasteries, these rights had supposedly reverted to the Crown. Surviving papers in this case, dating from about 1571, include a list of questions to be put on behalf of the Crown's representative as defendant. Answers given by the defendant's witnesses state that Barnardiston had no boat, which would be needed to fish, while the College kept one in its boathouse. A key witness was the Archbishop of Canterbury, Matthew Parker, who gave evidence as a former Dean of Stoke College. His testimony recalls that the matter was discussed when he dined with Mr 'Barmston' on 2 June 1543, so this was an issue which had caused difficulties for decades.

The mapmaker was John Hunt, a commissioner in the case, and related to Sir Ambrose Cave, who was involved with the drawing of Knaresborough Castle. Hunt explained in a signed note that he and the plaintiff's friend, Thomas Lark, differed on 'some points of small weight', so Hunt made this map and Lark made another. This second map is not known to survive, so we cannot compare the two, to judge the truth of Hunt's claim that the differences were 'not much material' to the case. We do have this map, which brings alive a summer scene in a corner of the country.

BOATMAN IN BREECHES: Depictions of humans are rare on early maps. This boatman among fish and swans may have been drawn because a boat was mentioned in court papers, as crucial to enable fishing on the lake.

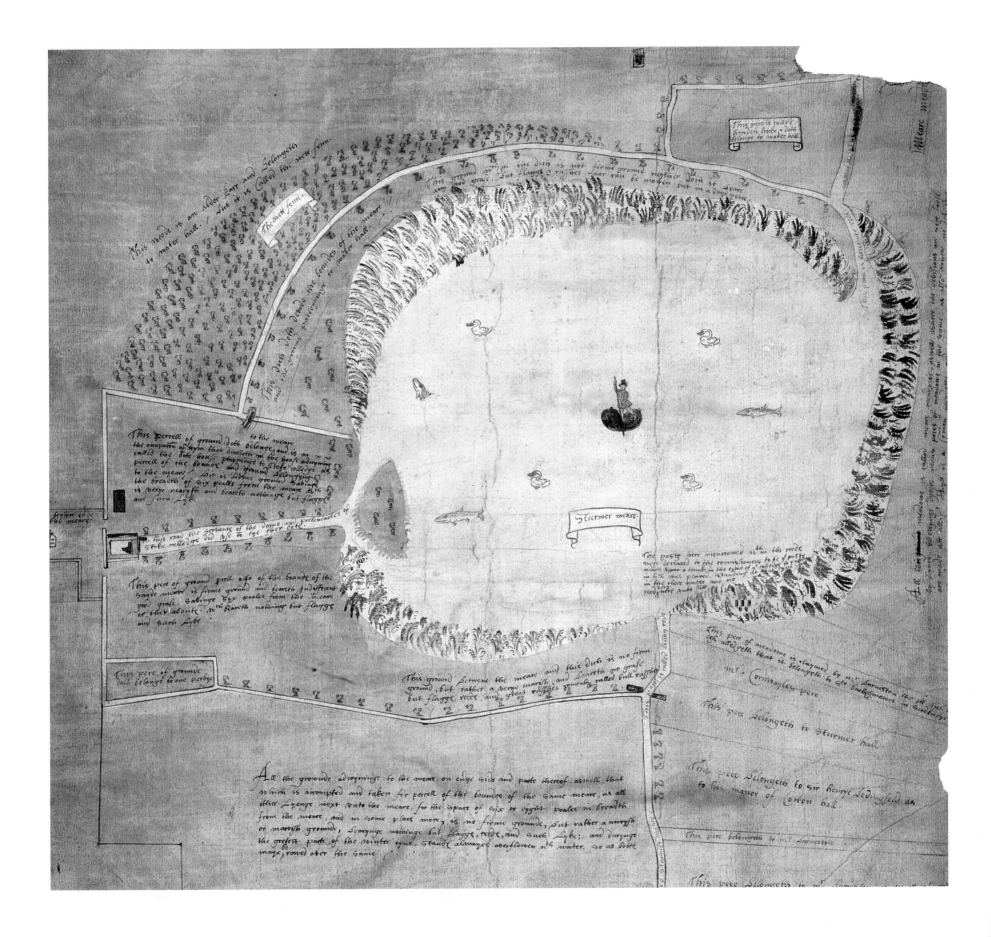

# Sea change: a receding coastline

Is this a map or a chart? On paper as in real life the line between land and sea has blurred edges. What it shows is the town and Cinque Port of Rye at centre, at a pivotal point in its history. Its heyday was the mid 16th century, but just decades later when this map was made its fortunes were beginning to wane. Ship traffic was much reduced from 1558 when England lost her last continental possession, Calais, across the Channel. And the waterways from the town to the sea (top) were beginning to silt up, beginning the process of stranding the port inland, miles from the sea, as now. This meant that ships had to unload goods into boats of lesser draught to reach the town. Rye was defended from the sea by Camber Castle (top centre), built by Henry VIII but becoming obsolete as the sea moved away from it, and abandoned a few decades after this map was drawn. Winchelsea town is in the top right corner, while the Playden village is represented below Rye by its church and mills drawn 'upside down'.

The map was made and signed by John Prouez or Prowse, an innkeeper of Rye. Like most Rye men, Prowse was also a mariner, which may explain why this map has the hallmarks of a chart of this date although it was not intended for use at sea. Four compass roses have large arrows pointing to north at the lower edge of the map, while radiating from them are rhumb lines, which were used by sailors for direction finding. The waters swarm with detailed drawings of different vessels, from sea-going men-of-war with open gunports and full rigging, to fishing boats by the town, and rowing boats to ferry men and goods between ships and the town. These boats are alive with sailors, though to the left of Rye are people in the water who seem to be bodies rather than simply out swimming.

The marshes and creeks around the town were the reason for this map's making. Below and to the right of the town a note indicates a breach in 'Mr Sheppards Marrshe'. This apparently relates to a petition to the Privy Council in 1572 by Robert and Alexander Sheppard from nearby Peasmarsh (off the map at lower right). They explained that a mighty storm in November that year had caused 'a breach by rage of water', and flooded an area which had previously been marsh, but had been drained or 'inned' to form a spit of land on which houses and mills had been built. They wished to repair the walls to protect these new buildings before further 'extremity of winds, weather and tide', but claimed that they were hindered by the townsfolk of Rye, who favoured a freer flow of water to feed the creek to their haven, to the right of the town. As this map shows, major coastline change is nothing new.

TUDOR TOWN: Rye's buildings including the central church were drawn in perspective within its walls, with gates onto the marshes. The mapmaker was keeper of the town's ordnance, and drew gun gardens bristling with cannons pointing seaward.

# Rabbits galore! METHWOLD WARREN, NORFOLK, 1580

Sixteen rabbits are drawn in silhouette on this map, one facing the opposite way from the rest. They are all found on the area coloured yellow, Methwold Warren, and represent many others of their kind. These 'conies', and their tendency in reality to stray beyond the warren, were the cause of many local disputes. The warrener paid a fee to the Duchy of Lancaster, who owned the manor of Methwold, to farm rabbits there, which he expected to recoup by sales of their meat and fur. He suspected his neighbours of trespass on the warren to take rabbits, but these neighbours said that they were merely dealing with rabbits that strayed on their land and ate their crops.

The map aimed to set out the boundaries of the warren so that no one could be in doubt as to where the rabbits were supposed to be. Before this time, local maps might be drawn by people such as gentry or clergy, with no particular expertise in making maps. In this case, one of the new professional surveyor-mapmakers was called in. Around 1580 instruction manuals were published, new instruments available, wealthy landowners willing to commission maps for estate management or as status symbols, and (as here) courts now requiring maps as evidence. We know little of John Lane, who made and signed this map, but he may have made two similar maps for disputes heard in the Duchy of Lancaster Court.

The map bears the hallmarks of the new style. It is set in a border with cardinal points in Latin, still in common use in documents of the time. West – Occidens – is at the top of the map. A large scale bar with dividers draws attention to the fact that the mapmaker used scale, a relatively new concept in these types of maps. From this we can work out that the warren was about four miles across at its widest, from north to south. So it was quite a long way for the party who, John Lane notes in the box at top left, walked around the edge of the warren to agree the boundaries. These men were commissioners appointed by the Duchy Court, who signed the map's top edge, and 'ancient inhabitants' of the area. A carefully drawn line of dots shows the route they took

The map shows how the landscape around the warren must have looked, especially features on the boundary such as hills, crosses and gravel pits. A raised bank is shown crossing the warren, dividing parishes. Churches drawn in perspective indicate surrounding villages. Near Methwold (top right) are two post windmills and 'Methold Lodge', which was recorded in 1413 and survived for another 27 years after this map was drawn. Lane tells us that the local elders viewed the finished map, and swore that the warren's boundaries had been as shown on it, for as long as they could remember. However, rabbits cannot read maps, and the disputes continued after the map was made.

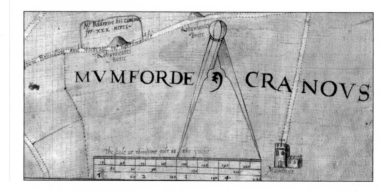

A QUESTION OF SCALE: John Lane wrote the scale in words above the scale bar, 'threescore poles to the ynche'; 1 inch to 60 poles. A pole, also called a rod or perch, was an old-style unit of land measurement, equal to five and a half yards. Despite this, he explained in a note that he had not drawn the surrounding area to scale, only the warren – and presumably not the rabbits!

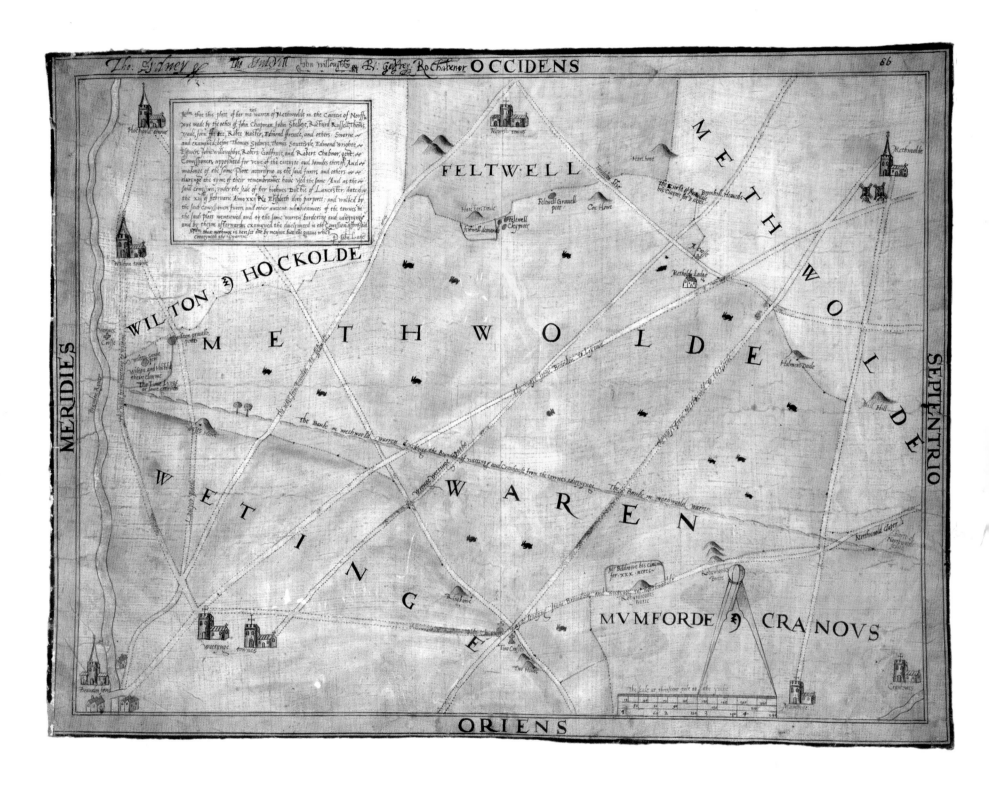

# The theatre of the sky and earth

If this map of the world appears to be dominated by a vision of purgatory, that was not the design of the mapmaker. We know from a surviving complete example, in Chicago's Newberry Library, that this map sheet was the lower of two, and that the upper sheet showed a wider view of the heavens. Text sheets joined on either side to form a broadside, a popular print format for ephemeral material such as posters and news. The non-map content is attributed to Giuseppe Rosaccio, a Venetian publisher who styled himself 'Doctor of philosophy and medicine, and cosmographer'. He published the text separately, as 'Teatro del cielo e della terra'; this is also the title on the upper sheet of the complete map. The use of Italian rather than the more scholarly Latin suggests an intended readership that was local and less well educated.

The cosmic and religious context is emphasised by large circles above the map, which contain calendars of Golden Numbers to calculate Easter and Epact Numbers to fix lunar years. These start in 1603 and could be used until 1621, so this was a product with a limited life expectancy. One of the calendars was made by Aluise Rosaccio, perhaps a relative of Giuseppe. The map itself was made, a note at the lower edge states, 'in casa del Signior Francesco Robacioli'. This house or workshop was in Brescia, a town in Lombardy, northern Italy, about 100 miles west of Venice. The address comes from the right-hand missing text sheet, dated 1602.

The map shows what was known of the world at that time. Places such as Tartary, Arabia Felix and Barbaria have the ring of antiquity to our ears. The outlines of countries are generalised and their placement approximate. North America, in so far as it was known, was then called New France. At left above the Tropic of Cancer lies the semi-mythical Strait of Anian, a supposed end of the long-sought Northwest Passage. Nothing was known above Greenland and Novaya Zemlya, north of Russia, or below Indonesia. Tierra del Fuego, the southern tip of South America, is shown as part of what was believed to be a vast southern continent across the lower edge of the map. Across this hypothetical space and the seas are drawn beasts, birds and sea creatures; a strange mixture of the dragon-like or mythical, and the simply wild.

This now very rare map was found among papers of Sir Joseph Williamson, Keeper of State Papers 1661–1702, a Secretary of State and an antiquarian. His collection was made from official records and from private sources, so it is difficult to say exactly why this map is in the archives. The collecting instinct appears to have by chance preserved this map down to our times. Its rarity reminds us that, just because a map was printed, even in some numbers, this did not necessarily ensure its survival.

PURGATORY PORTRAYED: Above the world map, this vivid view shows the journey of souls delivered by the ferryman to the Inferno to be cleansed by fire, thence to the next circle Limbo, where, wrapped as if mummified, they await Judgement Day.

42

# Around France

Intelligence, in the sense of information useful for diplomacy, has been gathered in all forms by governments for centuries. There were a number of foreign-made maps in the War Office, now held in the archives, which appear to have been collected simply because they showed areas which might be of military or political interest. This is one of a handful of 17th-century French maps from this source. It was made during the Thirty Years War which involved France and much of central Europe, and three years before the birth of the prince who would soon become the Sun King, Louis XIV.

This circular map of part of western France has at its centre the town of Niort on the River Sèvre about half way between Poitiers and La Rochelle in the Poitou-Charentes region. There seems to have been a practical reason for the shape of the map. Eleven concentric circles radiate from Niort, each representing a length of one French league (about 2.5 miles), and this would have enabled the ready calculation of distances. The map spills over beyond the outermost circle at top right to the River Charente and the town of Chateauneuf.

The map was apparently administrative in purpose, since the colours show divisions of the country into 'élections', larger fiscal districts about the size of a county. It also refers to 'châtellanies', areas under the governorship of a local castle. A note at one edge records the transfer of 54 parishes within three châtellanies from the élection of Niort to the élection of Cognac. This note is dated 1636, so the map was presumably made then.

The map was signed as having been 'painted' by F Granier. It is a pen-and-ink drawing on parchment with watercolour used to highlight certain regions and to embellish details. Villages and towns are represented by perspective drawings of churches, or of buildings around the churches. The Charente, the Sèvre (with a ship in its estuary), and other rivers are shown coloured blue. Large bog plants reflected in the water indicate the marshes of the Marais, between the pink and green areas. Other features include forests, principal roads and bridges drawn from life. These would all have been of interest to any commander of an invading army.

On the west coast, at the lower right edge of the map just outside the outer circle, lies the port of La Rochelle. Two towers guard the harbour and the town is shown in some detail. English military commanders might have found this view useful had it been in their hands in 1628, when they tried to relieve the long siege of the town's Huguenot population by Cardinal Richelieu on behalf of Louis XIII, the Catholic ruler of France. The English fleet sent to help them by Charles I failed to break through the French fortifications, not shown on this map, and with no other Protestant power to help, the defenders surrendered.

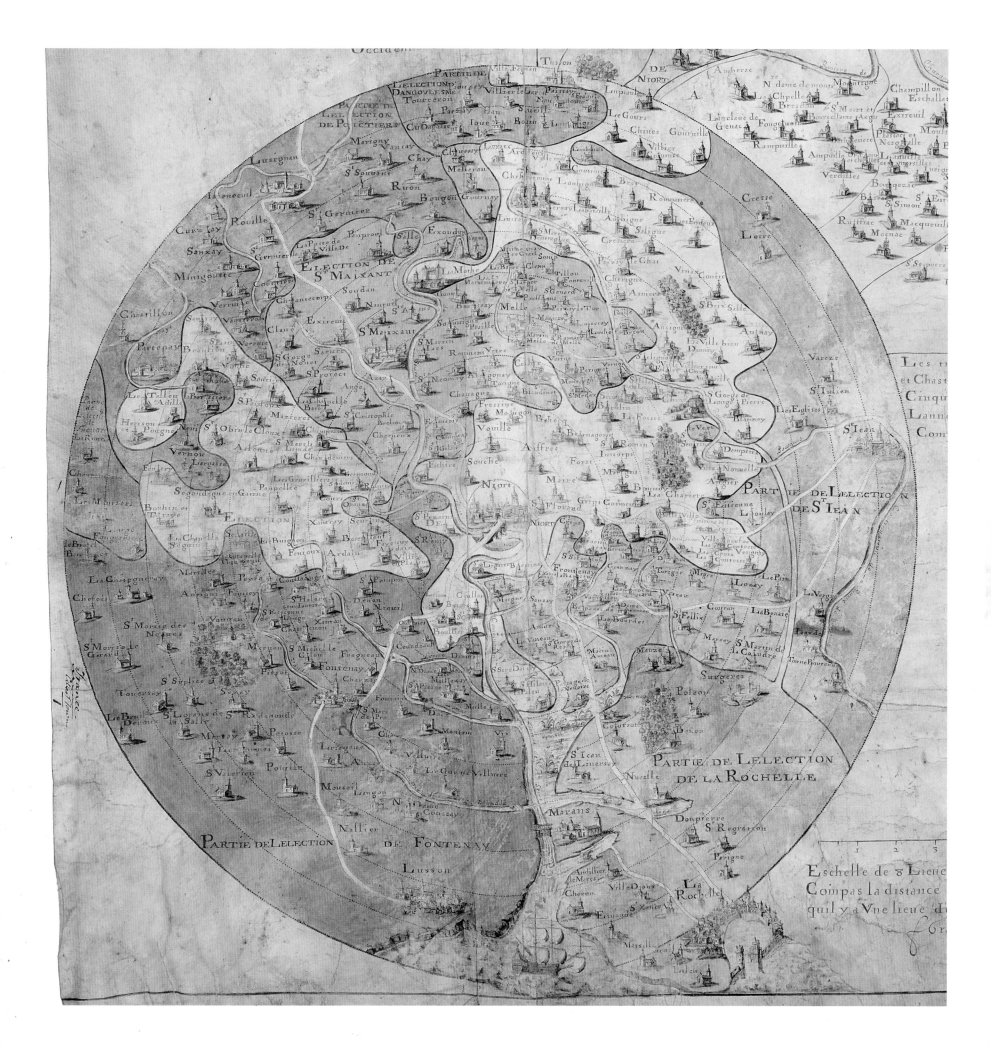

# CHAPTER TWO

# MAPPING THE metropolis

For thousands of years, people have striven to adapt and transform the natural world to create environments where they can thrive and live comfortably. Of all the habitats that humanity has devised, perhaps the most extreme – and the most profoundly human – is that of the city. Archaeologists have discovered that urban living dates back to at least 4,000 BC in some parts of the world. Over the past few centuries, it has become steadily more popular. City-dwellers now make up about half of the world's population, and this proportion continues to increase.

Cartographers have often found the city a difficult place to map. It is in the nature of urban environments that land-use is both dense and diverse, with buildings of different ages and purposes frequently sited in close proximity. Rapid changes are the norm in many cities, especially those with growing populations to accommodate. In others, buildings and localities of special historical or cultural interest are zealously protected from alteration. In fact, the world's urban areas are so

diverse that few generalisations about them can be wholly accurate. They vary enormously in age, in size, and in political and economic importance. Their terrain may be flat or hilly, their thoroughfares broad and straight or narrow and twisting, and their development laissez-faire or tightly controlled.

The maps of cities held among the records at The National Archives are no less diverse than the urban landscapes that they portray. Since the last years of the 16th century, the British government has made significant use of both printed and manuscript mapping of urban centres. Indeed, many of the oldest and most beautiful of our printed maps are town or city plans. By the 19th and 20th centuries, maps in the government's custody feature cities from almost every part of the globe, as well as towns throughout the British Isles.

Inevitably, the city most frequently represented on maps in the archives is London, the capital of the United Kingdom and the former hub of the British Empire. As the core of this chapter, we have therefore selected three maps capturing

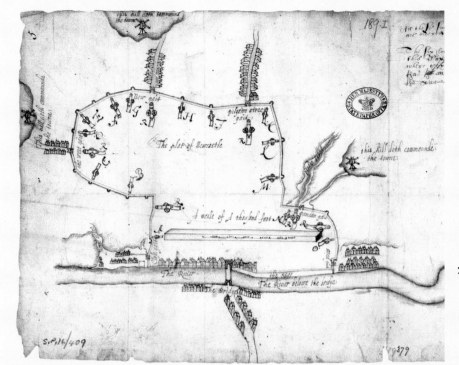

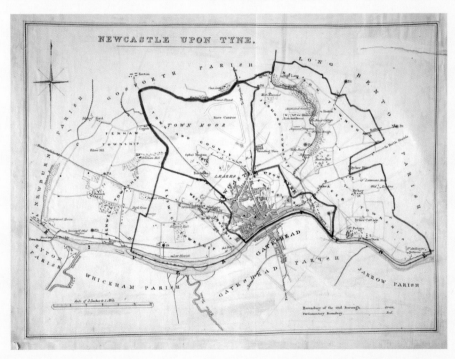

**1  NEWCASTLE UPON TYNE:** These two plans show the north-east of England's chief urban centre as it appeared in the 1630s and 1830s. The town was awarded city status by Queen Victoria in 1882.

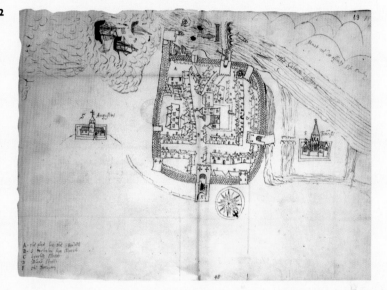

**2  GALWAY, IRELAND:** Drawn in 1583, this manuscript town plan is full of lively detail.

**3  JERUSALEM:** This plan dating from 1838 depicts Jewish, Muslim and Christian sites in one of the world's most important religious centres.

this great metropolis at moments roughly 140 years apart. Each was made at a time when maps were needed to record changes in the urban landscape. Restoration London of the 1660s (see page 53) and Georgian London – depicted at the time of the Napoleonic Wars (page 63) – were both encroaching inexorably on the surrounding countryside. By contrast, London in the autumn of 1940 (page 75) was struggling valiantly for survival against the chaos of the Blitz. Excluded from this chapter due to lack of space are maps of London during the Victorian and Edwardian eras, when it was undeniably the greatest city on Earth; we have sought to rectify this omission in Chapter 8.

To complement these portrayals of London, we have chosen maps depicting ten other cities from across the world, from mighty New York (page 61) to tiny Willemstad in Curaçao (page 55). These offer a taste not only of the broad chronological and geographical coverage of urban maps within the archives, but also of the varied purposes for which they were created and used. Our earlier examples tend to reflect two concerns: defence and civic pride. For instance, both fortifications and coats of arms feature prominently on the maps of Valletta (page 51) and Frankfurt am Main (page 57). In the second half of the chapter, concerns with security and status begin to give way to the practicalities of urban development and administration. The maps of Adelaide (page 67) and Nairobi (page 73) address town planning, while the regulation of commerce inspired the map of Cape Town (page 65).

Several themes persist across the continents and centuries covered by these city maps. One of these is the careful rendering of streets and buildings that define urban landscapes. Another is a preoccupation with water. Many successful cities are also ports, and several of these maps feature sailing ships as decoration. Something common to all of the maps is that they fulfil the need to make sense of, and assert control over, the complex and sometimes unruly world of the metropolis. In this chapter, we invite you to explore the urban jungle through our maps.

**4**

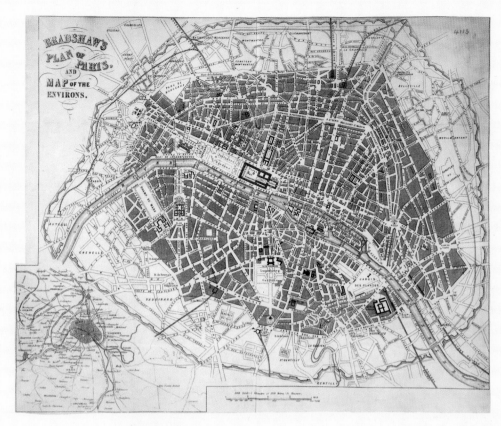

**4** PARIS, FRANCE: Often seen as a rival to London, the French capital has experienced an equally turbulent history. This British-made map showing the city's burgeoning railway network was printed in 1850.

**5** SINGAPORE: This modern-day city state was founded by Sir Stamford Raffles in 1819 and became an independent country in 1965. It is shown here in 1949.

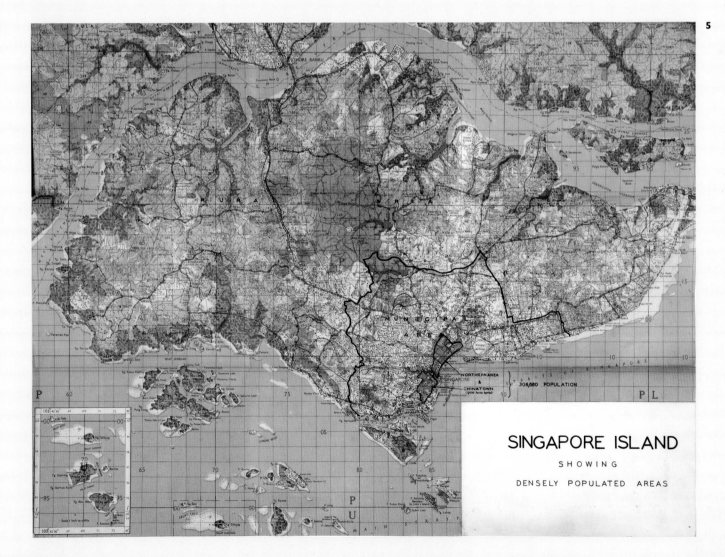

**5**

SINGAPORE ISLAND

SHOWING

DENSELY POPULATED AREAS

# The Knights of St John  VALLETTA, MALTA, c.1635–1643

The history of Malta has been shaped by its strategic location in the centre of the Mediterranean Sea. Between 1530 and 1798 the archipelago was controlled by the Knights Hospitallers, an international religious and military order of Catholic noblemen dedicated to St John the Baptist. They came to Malta at the invitation of the Holy Roman Emperor Charles V, after Ottoman Turkish forces had ejected them from their previous headquarters on Rhodes. In 1565 the Ottomans also challenged the Hospitallers' control of Malta. After a four-month siege with heavy casualties on both sides, the Knights and native Maltese – with assistance from Spain and Sicily – saw off the invaders.

Following this victory, the Hospitallers decided to build themselves a new capital on the Xiberras peninsula. Lying between two natural harbours on the north-east coast of the main island (also called Malta), this site made more military and economic sense than the ancient inland capital of Mdina. The Knights named their new city Valletta after their leader, Grand Master Jean Parisot de Valette. The city was designed on a grid plan, with strong fortifications to ensure its defence. Within a few decades, it had become a flourishing miniature metropolis.

This delicately hand-coloured engraved map is oriented with south-west at the top. Valletta occupies the right-hand side. To our left lie the 'three cities' that form part of Malta's formidable defence works: Bormla (or Cospicua) in the upper left corner, and the two peninsulas of L-Isla (Senglea) and Birgu (Vittoriosa). These are separated from the capital by the Grand Harbour. Several ships – two of which are flying the red and white flag of St John – plough through the water. The cartouche at lower left depicts Malta alongside its sister islands of Comino and Gozo.

Most maps among the archives that show places in Malta have a connection to the islands' former status as a British colony between 1800 and 1964, but this is an exception. Its origin and context are French. It is thought to have been prepared to illustrate a book by Brother Anne de Naberat, a French Hospitallers whose coat of arms appears below centre. Our copy is bound within a handsome volume of almost 300 plans and views of European towns and cities, many of which come from another publication, Sébastien de Pontault de Beaulieu's *The glorious conquests of Louis the Great*. The draughtsman may have been Daniel Rabel, an artist best known for his botanical illustrations and designs for French court ballet costumes. The engraver, Isaac Briot, produced an equally varied output, from portraits to coins and medals.

The map's text, however, is written not in French but in Italian. This reflects the fact that Rabel and Briot based their map on one made in 1602 by the Florentine Hospitaller Brother Francesco dell' Antella. This, in turn, was inspired by even older maps. Despite its small size and recent origin, the 'citta nova' ('new city') of Valletta was already the subject of an established cartographic tradition.

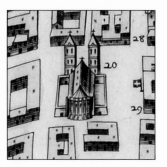

FORTRESS CHURCH: Built in the 1570s for the Knights Hospitallers, the church of St John the Baptist became celebrated for the contrast between its plain exterior and highly-decorated interior. It is now one of Malta's two co-cathedrals.

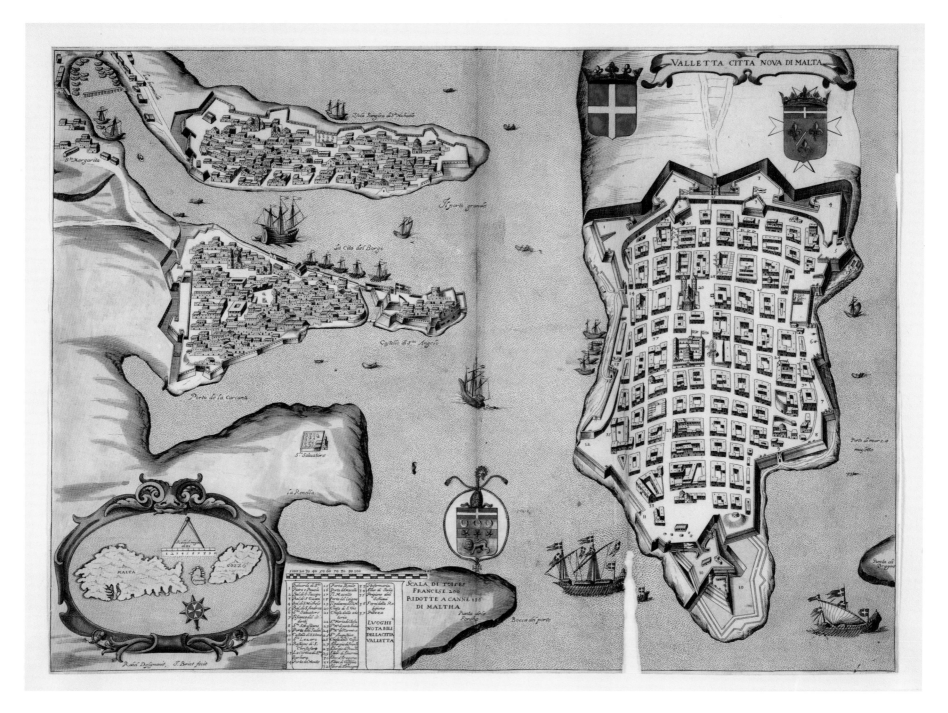

# Old Father Thames

LONDON, 1662

The name of the River Thames is inextricably linked with that of London, by far the largest settlement along its 200-mile course from the Cotswolds to the North Sea. Known to Londoners simply as 'the River', the Thames has for centuries played many central roles in the life of the city. At the time when this map was made in 1662, the river served as a county boundary, a vital transport link and the city's chief sewer, but no longer its only source of drinking water. Only the western end of the map is shown here, covering London and its immediate surroundings. The complete map is three times as long and depicts the Thames between the capital and its estuary in the east.

The map was made to fulfil a government commission. The man whom the Surveyor General, Sir Charles Harbord, selected to create it was Jonas Moore, who had mapped the Fens a few years earlier (see page 83). A leading mathematician, Moore took a thorough and scientific approach to his task. This is reflected symbolically in the dividers surmounting the scale bar and practically in the accurate detail of the city's streets. London was at this time enjoying a cultural and scholarly revival. King Charles II's assumption of the throne in 1660 – known as the Restoration – had marked the end of two unhappy decades in the nation's history, and the newly-founded Royal Society was promoting science and an understanding of the natural world.

More a work of art than a practical tool, the map is a fitting representation of a proud and confident metropolis. A lavish production in ink and watercolour, with liberal use of gold paint on the borders, it must have been breathtakingly impressive when new. It incorporates detailed scenic views – possibly the work of the artist and engraver Wenceslaus Hollar – of places along the river's lower reaches. The map's original owner was the Navy Board, so it is unsurprising that the places featured in these views, such as Greenwich, all have strong naval links.

Early users of the map would have included Samuel Pepys, who worked for the Board and was a close associate of Moore. Pepys would record in his diary the catastrophes suffered by London not long after the map was completed. In 1665, the city's population was decimated by a plague that killed more than 70,000 people. The following year, vast swathes of its buildings were burned to the ground in the Great Fire, which lasted for four days.

The map proved almost as vulnerable as its subject. It was kept for many years at the Office of Works, where it is thought to have been placed on display. Prolonged exposure to light probably contributed to the fading that has softened much of the detail over time, and made some of the writing difficult to read. Yet despite its now delicate state, the map survives as a visual record of a London lost to the changing fortunes of time.

RESTORATION LONDON: This panoramic view shows the metropolis from the south bank of the River Thames looking north.

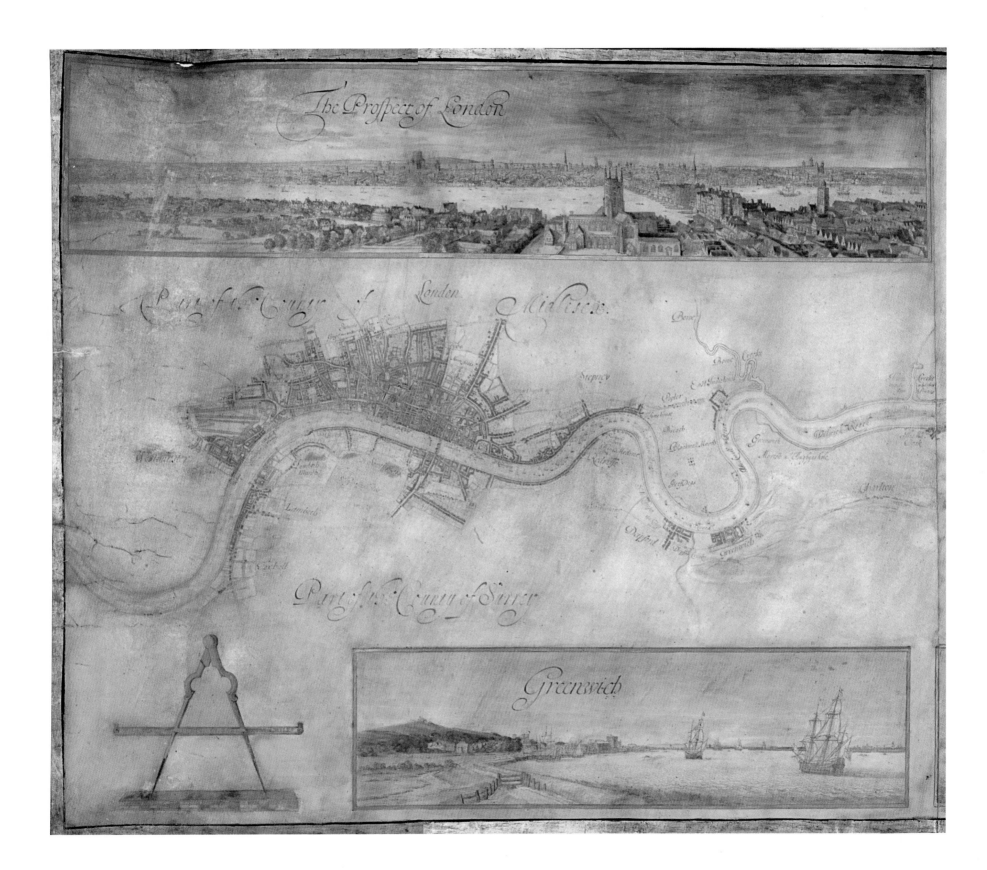

The Prospect of London

Greenwich

# Out of proportion

This intriguing item is really two maps in one. Its lower half depicts the Caribbean island of Curaçao, formerly a Dutch colony and now a country within the Kingdom of the Netherlands. The sliver of land at the top edge is the coast of Venezuela on the South American mainland. The red-edged shape occupying much of the upper right-hand side is not a second island but a large inset town plan. This shows Willemstad, the island's capital, drawn at 'ten times ye bignes' of the main map.

We know little for certain about when and why the map was created. It once belonged to the Board of Trade and Plantations and was later kept in the map library at the Colonial Office, where staff tentatively dated it to about 1700. We now think that it could be rather older. It may have been drawn between 1665 and 1675, during the Second or Third Anglo-Dutch Wars when military intelligence about Dutch colonies would have been of value to the English. At one point, perhaps when it was new, the map was stored folded in half; some of the ink has rubbed off onto the other side of the fold line, leaving behind a ghostly mirror image.

Compared to the detailed maps of London on the previous page and of Frankfurt overleaf, this depiction of Curaçao is almost shockingly crude. The coastline shown here is only a very rough approximation of the island's true shape and orientation; the proportions of the two large bays in particular are grossly exaggerated. Details of the interior are restricted to wavy green and brown lines, giving the vague impression of a rugged landscape. The scale bar at right and carefully-drawn compass indicator imply an attempt at accuracy and precision, although both are labelled upside down in comparison with the rest of the map. The stated scale is also inaccurate: Curaçao is actually about four times larger than it suggests. A second attempt at a scale bar, near the compass indicator, is unfinished.

Willemstad was founded after the Dutch captured Curaçao from the Spanish in 1634. At first the town was confined to a small promontory on the east side of the narrow St Anna Bay, at the entrance to the island's chief harbour. Today this area forms the city's historic district of Punda, meaning 'The Point'. The square outline enclosing three neat rows of houses may not be completely accurate, but it readily conveys how small the settlement was during its first few decades. Only in 1707 did Willemstad expand westward to the other side of the bay. The other enclosure on this peninsula is Fort Amsterdam, constructed to defend the new colony. Outside its defensive walls lie a collection of huts for African slaves. For today's viewer, these are an uncomfortable reminder that European economic success in the Caribbean was founded on the lucrative trade in human beings.

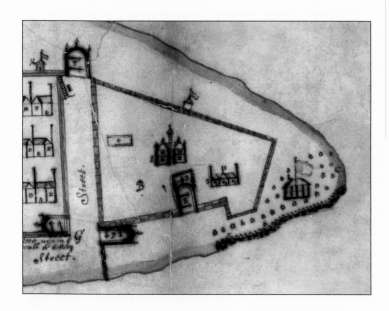

INSIDE THE WALLS: Both the town's church (numbered 1) and the governor's house (2) were sited safely within Fort Amsterdam. The distinctive colonial Dutch architecture of these and many of Willemstad's other historic buildings remains a prominent and colourful feature of the modern inner city.

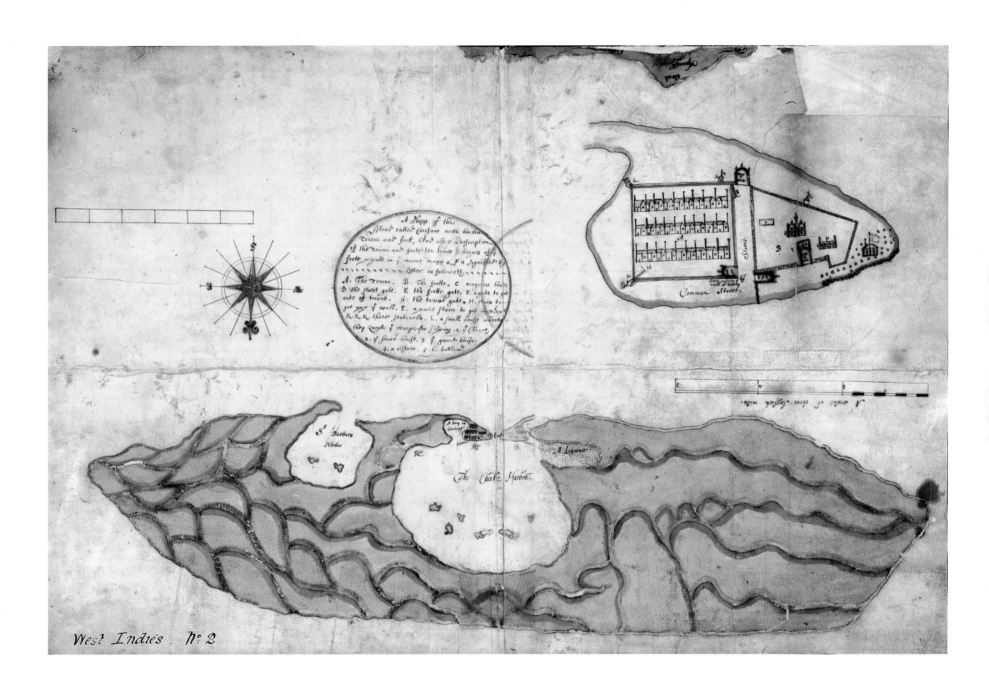

A Mapp of the
Iland called Curisaw with harbor
Towne and fort, And all a description
of the towne and fort ten times y⁷ bignes of y⁷
forte signifie in y⁷ maine mape & y⁷ is signified by
Letter as followeth

A: The Towne. B. The forte. C a narrow
D The streete gate. E the forte gate. F a gate to get
out of towne. G the towne gate. H stairs to
get up y⁷ wall. I a vast stair to get
K.K.K. theere sentonells. L. a small house w⁷ein
they supple y⁷ warp... &c.
... 2. y⁷ fourt house. 3. y⁷ guard house.
4. a cistern. 5. L. battlem⁷

S

N

Common Street

Street

St Barbers Harbor

R⁰

A Bay to Careene

A Lagoona

The Cheife Harbor

West Indies. No 2

# At home on the Main

In 17th-century Europe, maps and other engravings were often produced, published and sold by family businesses. Some of these small firms lasted for generations, passed from fathers to sons or other relatives. The creator of this map, Matthäus Merian the elder (1593–1650), was born into one such family – the Merians of Basel, in Switzerland – and married into another – the de Brys of Frankfurt. He learnt the art of engraving in Zurich and worked in several European cities before meeting and marrying Maria Magdalena de Bry whilst employed by her father, Johann Theodore. On his father-in-law's death in 1623, Merian settled permanently in Frankfurt to run the de Brys' publishing business, and later passed it on to his own sons. One of his daughters, Maria Sibylla, broke with the conventions of her era to achieve some success as a naturalist and scientific illustrator.

Merian is best known for his distinctive plans and views of towns. This map of his adopted home city was dedicated to the municipal government and originally published in 1628. Several later editions were produced, with revisions to reflect changes in the city's buildings. This version – issued three decades after Merian's death but still bearing his name – has been updated to 1682. It was engraved and printed as four separate sheets, which have been glued together to form the complete map. Our copy is accompanied by a short pamphlet containing a key to the numbered buildings marked on the map, and a brief history of the city, beginning with the 8th century Frankish King Pepin the Short and ending in 1682 with a public celebration marking the birth of a son to the Emperor Leopold I.

Frankfurt is depicted in bird's-eye view, from a slight angle. For the viewer, this creates a feeling similar to that which we now experience when looking down from a very tall building or a low-flying aircraft. We see the bustling metropolis that was the Merian family's familiar, everyday environment. Tiny people walk and ride through the streets of skilfully-rendered buildings; other citizens are sailing along the River Main or crossing the bridge on their way to and from the southern district of Sachsenhausen.

During this period, Frankfurt was part of the Holy Roman Empire, a complex agglomeration of states stretching across much of what are now Germany, Austria and northern Italy. It was proud of its status as a free imperial city, which made it answerable to no overlord except the emperor himself and gave it direct representation in the diet (or parliament). This pride is reflected in the cartouche at top right, which combines the imperial and municipal coats of arms. Frankfurt todayremains an important commercial and transport hub, and mainland Europe's leading financial centre. Over the past few centuries, its urban area has grown exponentially, leaving the walled city on this map – now known at the Altstadt, or Old Town – as a small district at its heart.

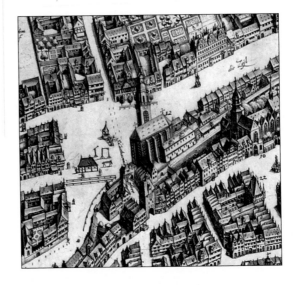

A NEW CHURCH: The Katharinenkirche, or church of St Katharine, was built between 1679 and 1681 to replace an earlier chapel. Like much of central Frankfurt, it was destroyed by Allied bombing in 1944, but later rebuilt.

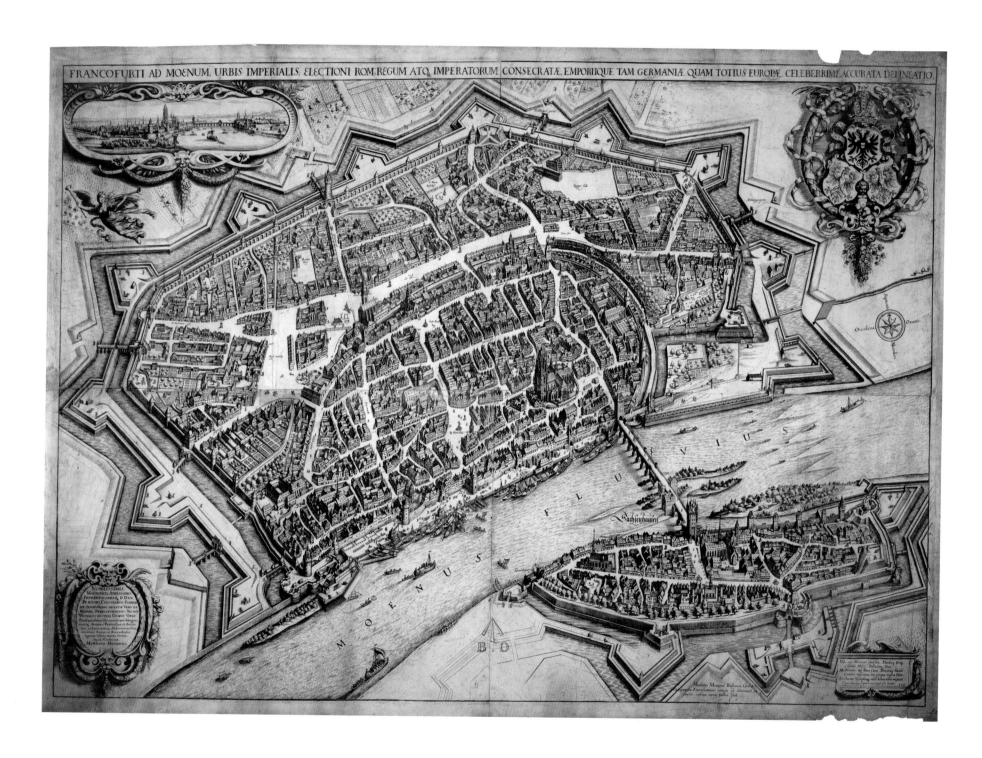

FRANCOFURTI AD MOENUM, URBIS IMPERIALIS, ELECTIONI ROM. REGUM ATQ. IMPERATORUM CONSECRATÆ, EMPORIIQUE TAM GERMANIÆ QUAM TOTIUS EUROPÆ CELEBERRIMI, ACCURATA DELINEATIO.

MOENUS FLUVIUS

Sachjenhauſen

# A ring of bright water

We can often learn a great deal about the meaning and purpose of a map from its context within the archives. Sometimes, however, we have little except the content of the map itself upon which to base our conclusions. This map of Mantua is an example of the latter kind. Our knowledge of its context is limited to the fact that it was kept by the War Office's librarians as part of a diverse set of manuscript plans of Italian towns and harbours, all thought to date from the 18th century. What practical use these would have been to the British government we can only speculate, but the variety of maps and plans accumulated by the War Office suggests that it maintained an interest in the defensive strengths and weaknesses of places throughout the world.

Military strategists during the 18th and 19th centuries must surely have found the distinctive urban layout of Mantua stimulating. The city has a beautiful lakeside setting on the fertile plains of Lombardy, in the north of Italy. Unlike Venice, which was founded with its canals, Mantua acquired its surrounding waterways part-way through its history. It was already an important city when the River Mincio was diverted at the end of the 12th century to engineer a defensive ring of four artificial lakes. The smallest of these, Lake Paiolo (meaning 'pot' or 'cauldron'), dried up in the late 18th century, leaving the walled city and the neighbouring island of Teieto permanently attached to the mainland. The other three lakes – Superiore, Mezzo and Inferiore ('upper', 'middle' and 'lower') – remain part of Mantua's unique topography to this day.

The city as depicted on this map superficially resembles an elegant castle encircled by a wide moat. The delicately stylised trees and very selective detail aid the false impression of a large fortress rather than a compact urban centre. In fact, a dense network of streets and buildings filled the space inside the city walls. The map is not drawn to scale; only the church buildings belonging to settlements surrounding Mantua indicate its true scope. Long causeways connect the main city to the outposts of San Giorgio (to the east, at top) and Porto Mantovano (to the north, at left), while shorter bridges allow access to the south and west.

The defensive lakes did not always protect Mantua from attack, notably in 1630 when it was sacked by the Holy Roman Emperor's troops, and in 1796–1797 when it was besieged by Napoleon. This map, however, is believed to date from a relatively stable period in the region's history. Like many places in northern Italy, Mantua long enjoyed independence, firstly as a city state and later as a duchy. After its Duke Ferdinando Carlo Gonzaga died in 1708, the territory came under the control of Austria. Apart from two brief periods of French rule, it remained part of the Austrian Empire until 1866, when it joined the newly-founded Kingdom of Italy.

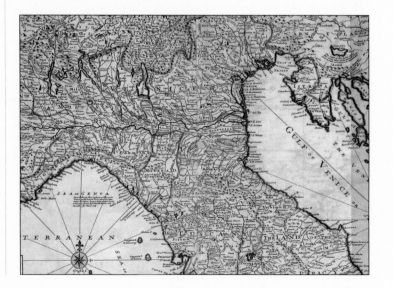

DUCHIES AND REPUBLICS: This engraved map by the London-based cartographer Herman Moll depicts the patchwork of states comprising northern Italy in the early 18th century.

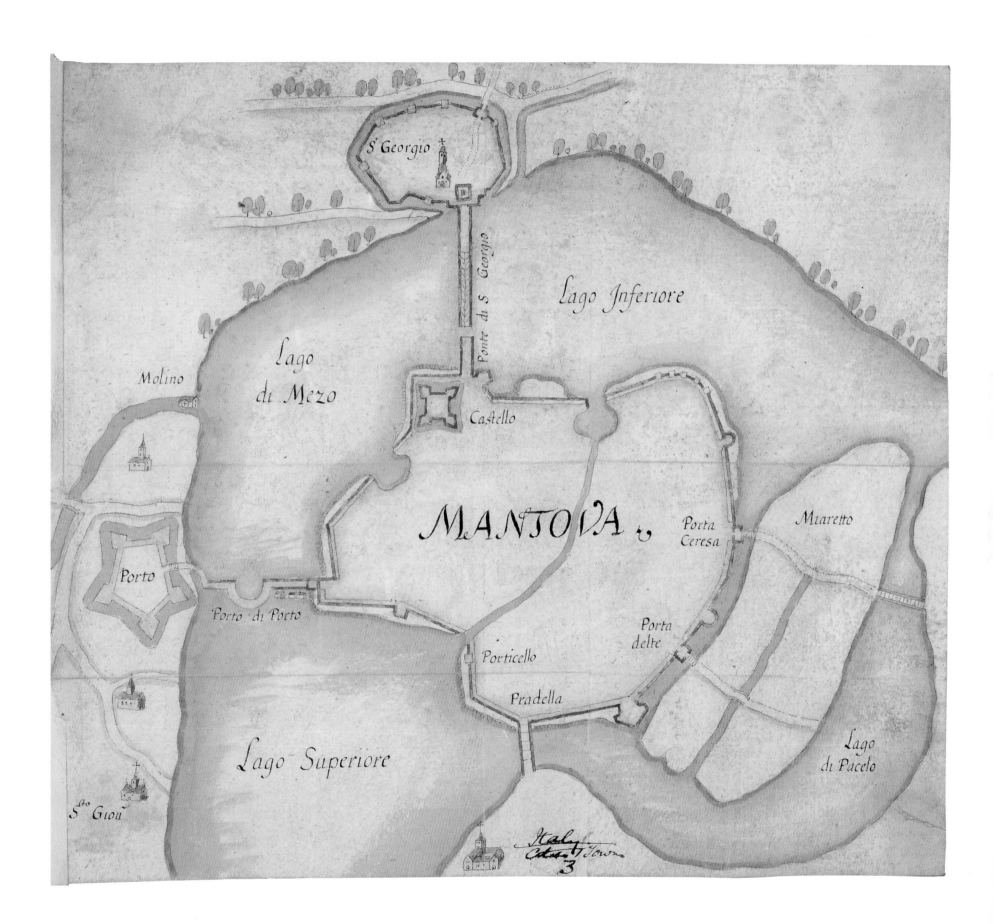

S.º Georgio

Lago Inferiore

Molino

Lago
di Mezo

Pontte di S. Georgio

Castello

MANTOVA

Porta
Ceresa

Miaretto

Porto

Porto di Porto

Porticello

Porta
delte

Pradella

Lago
di Pacelo

Lago Superiore

S.º Giou

Italy
Cities Town
3

# I saw three ships   NEW YORK CITY, 1765

Under the Stamp Act of 1765, residents of the British colonies in North America were required to pay a tax on many written or printed materials, including legal documents, newspapers and playing cards. Special paper bearing revenue stamps was imported from London and had to be purchased with English pounds, which were scarce in the colonies, instead of the ordinary colonial currency. The purpose of this stamp tax was to pay for troops to be stationed in North America following the British victory in the recent Seven Years War (or French and Indian War).

The Stamp Act was deeply unpopular among colonial residents, who decried it as 'taxation without representation', since the colonies did not elect representatives to Parliament in London. (Many British commentators at the time disagreed, pointing out that relatively few men – and no women – resident in the mother country could vote either.) There were widespread protests, including a riot in New York City that began on 1 November and lasted for four days.

This topographical view shows the city at the time of the riot. East is at the top, placing the East River (labelled K) furthest from the viewer and the North River (N) – an old name for the Hudson River – in the foreground. In the mid 18th century, New York City was a relatively small settlement at the southern tip of Manhattan Island. The very tall buildings that are characteristic of today's metropolis would not start to be built for more than a

century. Nonetheless, several features are easily recognisable. For instance, Bowling Green (labelled D) still lies at the southern end of Broadway and the Battery (B) is now the site of Battery Park.

The view was drawn by William Cockburn on behalf of Commander Archibald Kennedy, a senior officer in the British Royal Navy tasked with protecting Fort George (labelled A). This fort, which was also the colonial administrative headquarters, held a large stock of the all-important stamped paper. Kennedy's men were unable to protect the fort from the rioters and the governor, Cadwallader Colden, decided to hand over the stamped paper to the Corporation of New York, which destroyed it.

In an attempt to prove that he had tried his best to defend Fort George and its contents, Kennedy sent a bundle of paperwork to the Admiralty in London that explained and justified his actions. This view was intended to illustrate and support his argument that he had stationed his three ships – the Coventry (labelled F), the Guarland (G) and the Hawke (H) – in the best places. The Admiralty accepted Kennedy's explanations and he was allowed to continue his naval career.

The British authorities eventually realised that the stamp tax had proved too difficult to collect and repealed the Stamp Act in 1766. The colonists' successful resistance to the tax was a contributing factor to the American Revolutionary War of 1775–1783 and the founding of the United States.

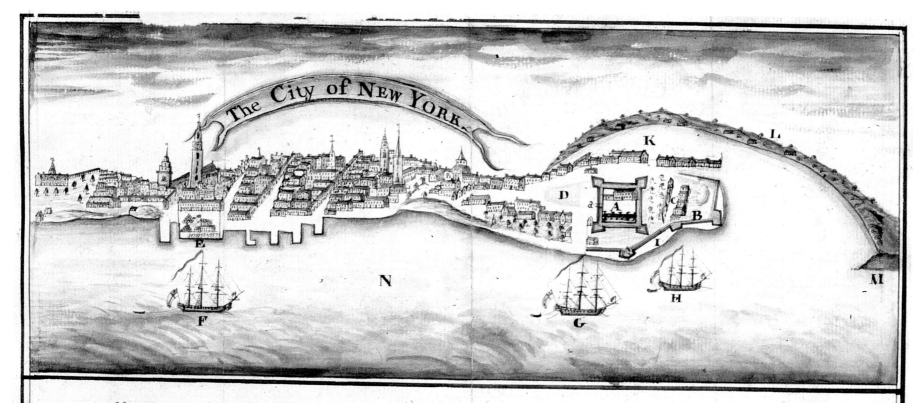

The City of New York

The Position of his Majesty's Ships as they where Stationed on the 1 day of November 1765.

A. Fort George, & The Fort gate. B. The Battery C. The Barracks. D. The Bowling green, and Broadway. E. The Kings Wharf, and Arsenal. F. His Majesty's Ship Coventry, to protect d.º G. The Guarland to scour the street, and defend the Fort gate, H. The Hawke to preserve a Communication between his Majesty's Ships & the Fort, by covering the landing of boats at I. The flat rock. K. The East River. L. Long Island M. Governors Island. N. The North River.

W Cockburn fecit

# Town and country

At the turn of the 19th century, London was a much smaller place than it is today. Many localities now considered central districts within the capital – for example, Knightsbridge and St Pancras – still lay outside the built-up area (tinted in light ochre on this map). A few miles further afield, villages like Clapham remained truly rural. The image opposite shows only part of a much larger manuscript map; the complete item extends much further to the north, east and south to cover the countryside as far away as Mill Hill, Barking and Croydon.

The pink blocks around the edges of this extract represent part of a ring of defences set up to protect London from an expected invasion by the French Emperor Napoleon. These provide a clue to the map's military origin. Drawn in February 1803 by W Chambers and J Anderson, two officer cadets at the Royal Military Academy in Woolwich, it was largely copied from an earlier map of September 1801, which is believed not to have survived. The map provides evidence of the degree of professional training that the British Army had begun to develop for men in technical roles, such as surveying and cartography, and the quality of its workmanship is exceptionally high. Hills are depicted vividly with hachures, offering a sense of the nature of the terrain that is unmatched by earlier maps of the region. Woodland, open spaces, roads, and even individual buildings are rendered in painstaking detail, ensuring that the map is of more than just military interest.

The map captures the landscape at a time when the social and economic ties between London and its environs were especially strong. The growing population of city-dwellers – nearly one million people by 1803 – relied heavily on the countryside for fresh food, and many rural farmers earned their living from customers in the capital. Meanwhile, members of the wealthy elite divided their time between residences in town and their country estates. London was also expanding rapidly. Ribbon development along major roads and early industries (such as gravel pits and brick kilns) on the fringes of the urban area formed a transitional zone between the countryside and the city.

In fact, drawing a sharp dividing line between London and its surroundings has never been easy. The distinction between town and country portrayed on this map reflects a common-sense understanding of its size and scope. Urban settlement had outgrown the tiny City of London and neighbouring Westminster – which was the seat of government and the royal court – long before the 19th century, but not until the creation of the Metropolitan Board of Works in 1855 would tentative steps be taken towards recognising this fact administratively. Since that time, most of the places shown on this map, and many areas beyond it, have been absorbed within the Greater London conurbation. These rural settlements of the past have evolved new identities as neighbourhoods, suburbs and 'urban villages' within the modern metropolis.

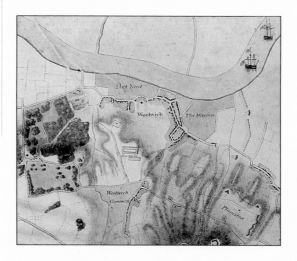

YOU ARE HERE: Woolwich, where this map was made, has had a long association with the armed forces. The Royal Military Academy was based here between 1741 and 1939.

# Fine wine

CAPE TOWN, CAPE COLONY (SOUTH AFRICA), 1833

Cape Town has a dramatic setting on the south-western coast of Africa. The city's historic centre sits between the waters of Table Bay and an impressive semi-circle of mountains, only the lowest of which, the Lion's Rump (now better known as Signal Hill) is shown here. By the early 19th century, the city's economic importance as a port and a convenient stopping point on long-distance voyages had made it by far the dominant European settlement in southern Africa. This map also records a strong military presence, particularly in the yellow-tinted area. Subsequent land reclamation has altered the shape of the coastline considerably, but the grid pattern of streets in the city centre remains largely intact. The many buildings shaded dark red are the premises of wine merchants, whose businesses had a long-established pedigree in Cape Town and a prominent place in its economy.

The first people to produce wine at the Cape were Dutch colonists of the late 17th century, who discovered that the area's Mediterranean climate was perfect for cultivating grapes. The arrival of French Huguenot immigrants with expertise in winemaking fostered the development of the fledgling industry. When the British took control of the Cape Colony during the Napoleonic Wars, they soon discovered the advantages of controlling a land where grapes were grown. Not since the English kings had lost their territories in south-western France in the 15th century had wealthy Britons enjoyed such secure access to a source of good wine. French imports were heavily taxed, and often unobtainable due to war between the two countries. Conversely,

customs duties favoured wines from Britain's ally, Portugal. A glass of port had become the traditional end to an upper-class Englishman's dinner.

Although import taxes on Cape wine were reduced to encourage trade with the mother country, the wine merchants of Cape Town remained frustrated by the practical difficulties of exporting their wares. The colony's governor, Sir Galbraith Lowry Cole, seems to have been sympathetic to the wine industry, since he revised the licensing and export regulations in its favour in 1832. What he could not do with his limited budget was to pay the £567 needed to build a new stone pier near Amsterdam Battery. This was desperately needed to replace the old and inconvenient jetty by the Castle of Good Hope. Cole wrote to London on 8 March 1833 to explain the situation, enclosing this map in his letter to illustrate the city's geography.

Despite such complaints, from traders and other workers employed in the industry, this period subsequently proved to have been a golden age for Cape wine. Three decades later, in the 1860s, the industry nearly collapsed under the triple impact of the oidium fungus, the phylloxera pest, and a trading agreement between the United Kingdom and France that ended preferential treatment for imports from the Cape. By the time that the wine trade recovered in the late 1880s, the balance of economic power in southern Africa had shifted to the interior, to the goldfields of the Witwatersrand and the new city of Johannesburg. Cape Town's era of dominance was over.

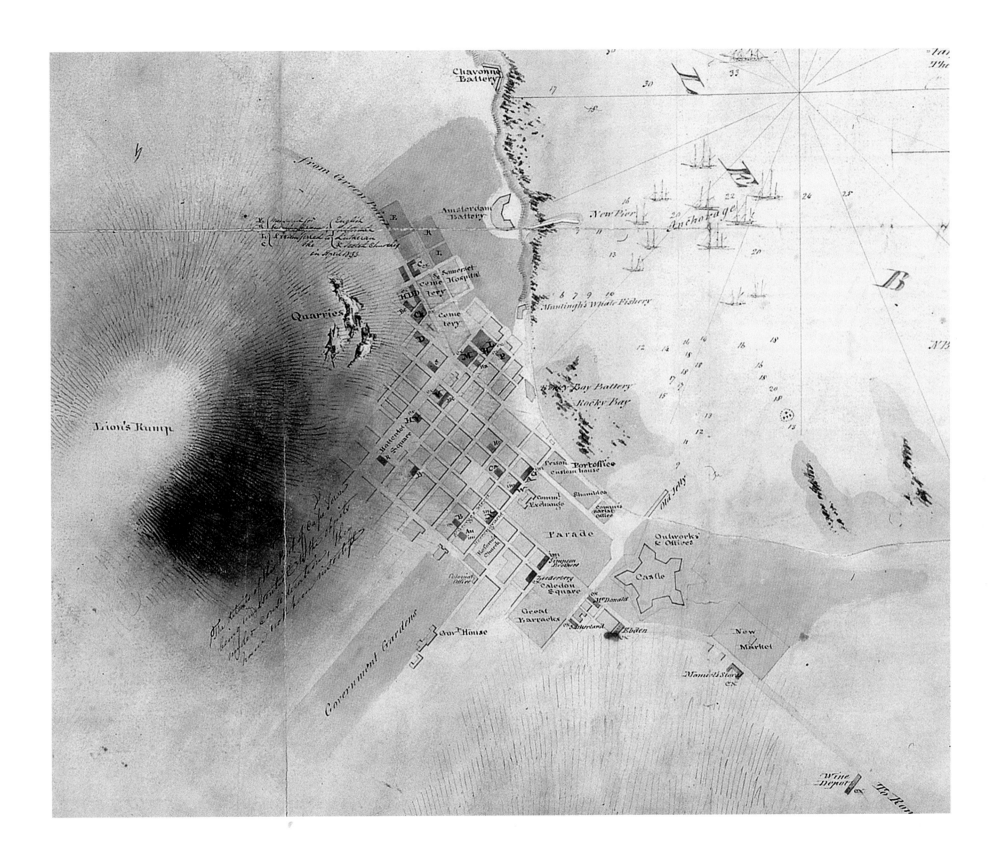

# A city fit for a queen

On 15 August 1834, the United Kingdom's Parliament passed the South Australia Colonisation Act, approving in principle the creation of a new colony on the other side of the world. This colony was not settled immediately. Only in February 1836 were the official letters patent issued in the name of King William IV to establish the Province of South Australia. Several ships of colonists set sail that same year, and the formal settlement of the colony was proclaimed at Holdfast Bay (near point H on this map) by its governor, John Hindmarsh, on 28 December.

Unlike some earlier British settlements on the Australian continent, South Australia was not founded as a penal colony for punishing criminals but as a community of free settlers. It was hoped that it would become a model of systematic colonisation, offering political and religious freedom to young men and women who were willing to work hard and establish new lives ten thousand miles from the country of their birth. Unusually, this emphasis on freedom was extended at least nominally to the region's Aboriginal people, whose pre-existence and rights were recognised in the founding documents.

William Light, a former officer in the British army, was appointed as the colony's surveyor general and tasked with planning its new capital. He chose a pleasant site (points B and C on this map) about six miles from Holdfast Bay. Light and his deputy, George Strickland Kingston, designed the town with a grid pattern of wide streets and squares, surrounded by parkland. Although Governor Hindmarsh was dissatisfied with Light's choice of site, which he believed was too far inland, his attempts to move the town to another location failed. The new capital was given the name Adelaide after King William's wife. This name is written rather faintly in pencil near the top right-hand corner of the map.

This sketch map of the area is drawn neatly to scale. It is covered with handwritten notes – many of which are quite difficult to read – outlining the nature of the terrain, and discussing the arrangements for landing ships and how they might be improved. The line between the points marked E and F marks the position of a proposed canal (which was never constructed), intended to link Adelaide with what later became Port Adelaide. The map is not signed and we cannot be certain whether Light drew it himself but it evidently reflects his opinions and intentions. It was probably drawn in February 1837 during the initial phase of surveying the town site.

In its early years, Adelaide suffered from economic difficulties and inadequate leadership but it soon began to thrive. With more than one million inhabitants, it is now the fifth most populous city in Australia and its metropolitan area stretches beyond the coverage of this map. Its historic centre still remains true to Light's original vision of a neatly planned town surrounded by a ring of green space.

CITY SQUARES: Although the scale of this map is too small to allow much detail to be shown within the area of the town, it does mark the positions of the five squares that still characterise central Adelaide. The central square is named after Queen Victoria, who succeeded William IV to the throne in June 1837. Two of the other squares are named after Light and Hindmarsh.

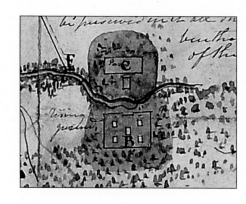

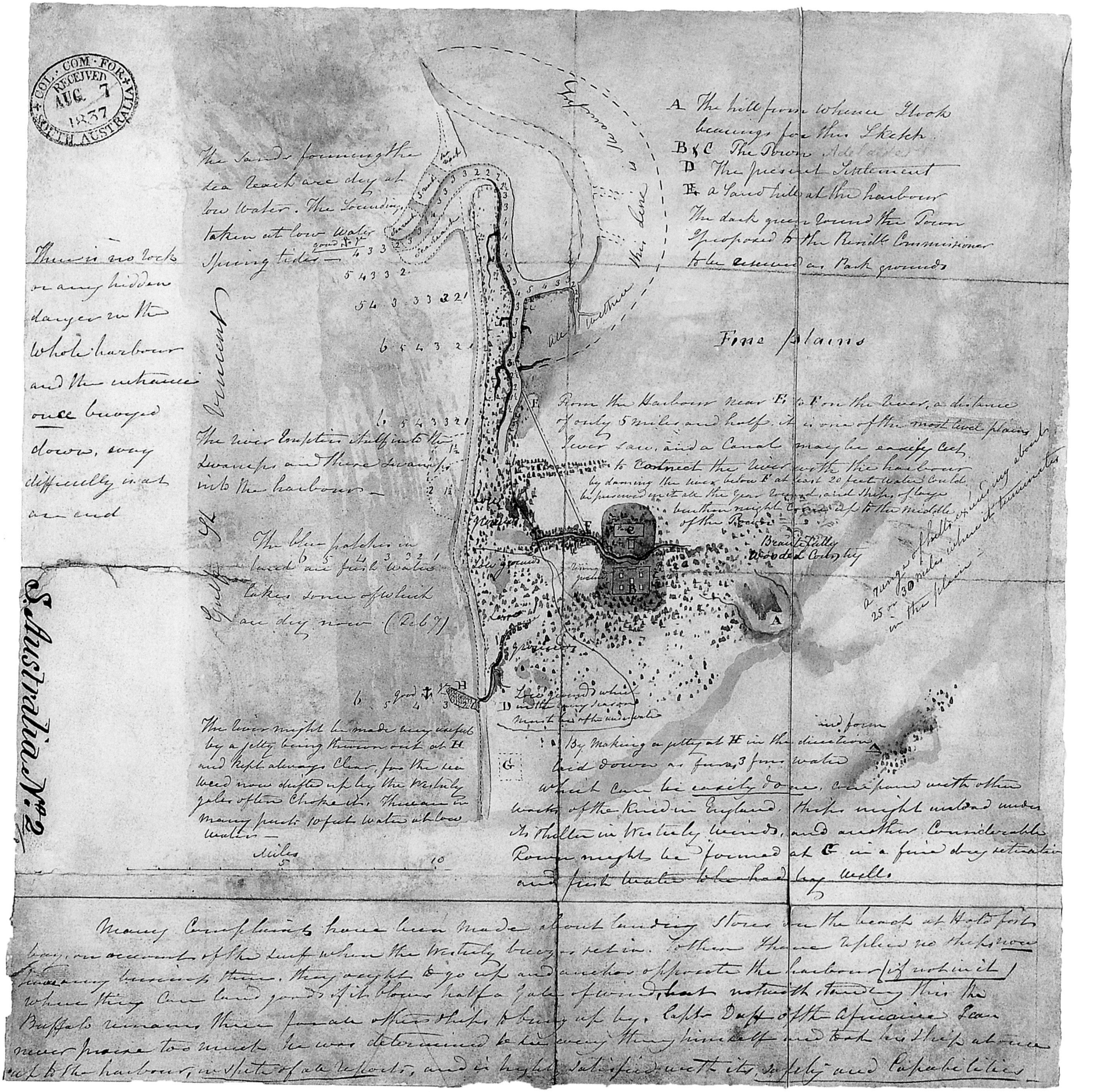

A The hill from whence I took
bearings for this Sketch
B & C The Town Adelaide
D The present Settlement
E a Sand hill at the harbour
The dark green round the Town
proposed to the Resident Commissioner
to be reserved as Park grounds

Fine plains

From the Harbour near E to F on the river, a distance
of only 5 miles and half, it is one of the most level plains
I ever saw, and a Canal may be easily cut
to connect the river with the harbour
by damming the river below E at least 20 feet water could
be preserved until all the year round, and Ships of large
burthen might come up to the middle
of the Reach

Beautifully
Wooded Country

The Sands forming the
sea reach are dry at
low water. The Sounding
taken at low water
Spring tides —

There is no rock
or any hidden
danger in the
whole harbour
and the entrance
once buoyed
down, very
difficultly in at
an end

S. Australia. No 2

The river Empties itself into the
Swamps and those Swamps
into the harbour —

The blue patches in
lined are fresh water
lakes some of which
are dry now (Feb. 7)

The river might be made very useful
by a jetty being thrown out at H
and kept always clear, for the sea
weed now drifts up by the Westerly
gales often choke it. There are in
many parts 10 feet water at low
water —

Miles

By making a jetty at E in the directions
laid down as far as 3 from water
which can be easily done, compared with other
works of the kind in England, Ships might unload under
its shelter in Westerly winds, and another considerable
Room might be formed at G in a fine dry situation
and fresh water to be had by wells

Many complaints have been made about landing stores on the beach at Holdfast
bay, on account of the surf when the Westerly breezes set in. I there upon no ships now
to continue landing there, they ought to go up and anchor opposite the harbour (if not in it)
where they can land goods if it blows half a gale of wind, but notwithstanding this the
Buffalo remains there for all other ships to bring up by. Captn Duff of the Africaine I can
never praise too much he was determined to try everything himself and took his ship at once
up to the harbour, in spite of all reports, and is highly satisfied with its safety and capabilities

# Eastern capital

EDO (TOKYO), c.1853

Tokyo today is one of the world's great metropolises and the core of one of the most densely populated urban areas on Earth. Yet, just a few centuries ago, it was a small, coastal settlement called Edo (or Yedo), meaning 'bay entrance'. Edo's rise to economic and political importance began around 1600 when the leaders of the Tokugawa shogunate decided to base their military government there. The ancient city of Kyoto remained the official capital of Japan and the seat of its Emperors, who at that time had a symbolic role without real power. When the Meiji Emperor regained political control from the shogunate in 1868, he moved his court to Edo, by that time a thriving city, and renamed it Tokyo, which means 'eastern capital'.

From the mid 17th century onwards, the shogunate had pursued an isolationist foreign policy known as sakoku. Economic links were very tightly controlled – for much of the time the Dutch East India Company held a monopoly on trade between Japan and Europe – and contact between Japanese people and foreigners was severely restricted. As the government began to relax the policy in the 1850s, western interest in Japanese art and culture grew rapidly.

This colourful map of Edo was made just at the time when Japanese society was starting to open to contact with Europeans and Americans. It is a woodcut made by three Japanese engravers: Manjiro Izumoji, Jibe Moriya and Kibe Wakabyashi. Centred on Edo Castle, it shows gardens, pagodas, temples and shrines, as well as streets and canals. The large block of text in the lower left-hand corner includes a list of festivals for the indigenous Japanese religion of Shinto. Many of the map's features look familiar to twenty-first-century western eyes. The use of colouring – yellow for streets, blue for water, green for parks or gardens and red for important buildings – gives it a superficial resemblance to some modern western street maps. The use of a ship in coastal waters as a decorative feature is a common trope on western maps, including several of the other maps in this book.

The map came to The National Archives as part of a journal kept by a Royal Navy surgeon named Charles Courtney. This is one of more than 800 surgeons' journals – many of which include maps – now preserved within the archives of the British Admiralty. We believe that Courtney acquired the map in about 1859 whilst serving on board *HMS Highflyer* during the Second Opium War with China. Courtney's journal also includes maps of several other places, including the Chinese cities of Amoy (Xiamen) and Canton (Guangzhou), illustrating his studies of the relationship between geography, climate and human health. Most maps and plans among British military and naval records were made or used by the armed services for their operational or strategic value. Charles Courtney's map of Edo is different: rather than having any practical function in the business of government, it was collected purely out of scholarly or cultural interest.

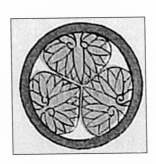

FAMILY CREST: This yellow floral symbol marking the site of Edo Castle is the crest of the Tokugawa family.

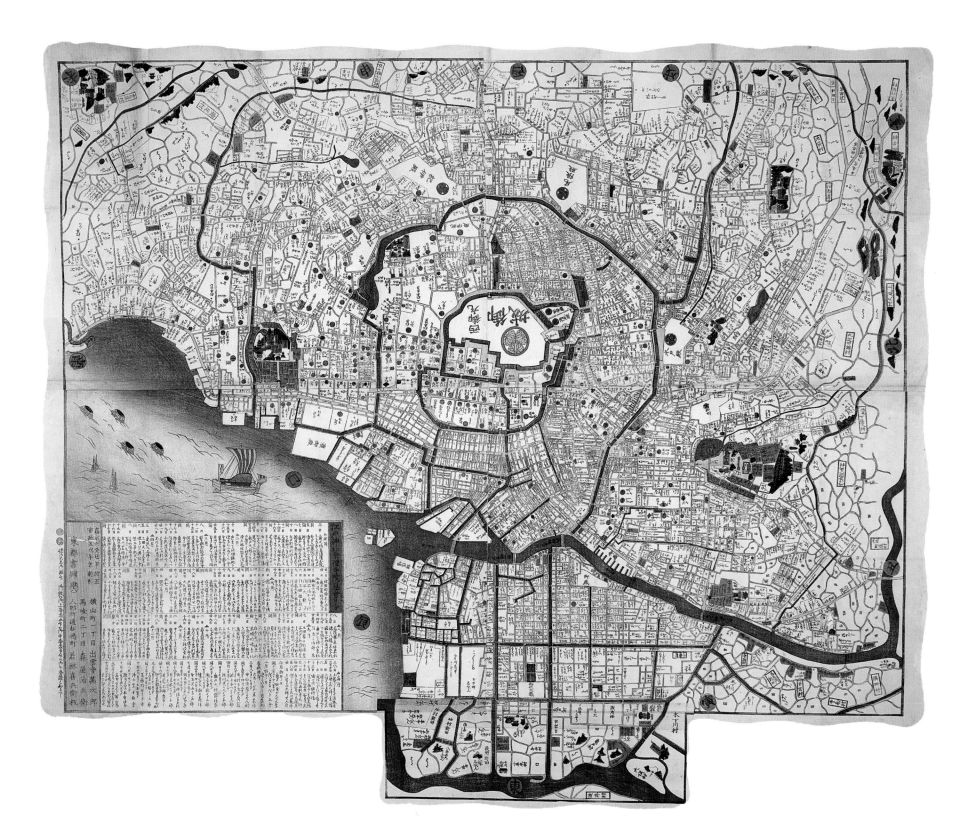

# The turtle and the missionaries

The capital of Sichuan province in south-western China, Chengdu has long been acknowledged as one of China's greatest cities. Urban settlement on its site dates back more than two millennia. For a long time, the city was especially notable for two walled districts within its boundaries: the rectangular Imperial City, shown near the centre of this Chinese map, and Shaocheng, the relatively uncoloured area lying immediately to the west. One of Chengdu's nicknames is Turtle City. Legend states that its outline was determined by following the route of a crawling turtle. It is also said to resemble the shape of a turtle when viewed from above, although that is not apparent from this map.

Western visitors to Chengdu have not always appreciated its rich heritage. In the early 20th century, foreign diplomatic staff regarded it as a backwater with an unhealthy climate, and significantly less prestigious than postings in eastern China. Accommodation suitable for consular use was not easy to obtain in the city. Westerners found Chinese houses uncomfortable, and opportunities to construct European-style buildings were rare. When the temporary home of the British Consulate-General was destroyed in a fire on 23 January 1906, its staff faced a daunting task to find new premises.

The Acting Consul-General, Herbert Goffe, used this map to identify potential locations, marking three suitable sites upon it in blue pencil. After several months of wrangling over the costs, the purchase of site A was agreed in August. Mr Goffe then encountered the further difficulty that China did not officially permit foreign governments to buy property from private vendors. Previously, diplomatic staff based in Chinese cities had circumvented this rule by using Christian missionaries (who were not subject to the same restrictions) as intermediaries; the

properties would be transferred into the Consul's name as soon as the sale was completed. Accordingly, Goffe asked a British missionary named Mr Edgar to make the transaction on his behalf.

Unfortunately, signs were erected by mistake stating that the property belonged to the British Protestant mission, instead of the Consulate-General. This upset Chengdu's American missionaries, who owned a neighbouring plot and had wanted this land for themselves. Hearing that the site had actually been acquired for diplomatic purposes, the Americans persuaded their Canadian and British colleagues to join in protest against the alleged exploitation of missionaries as political agents. The Chengdu authorities refused permission to register the land in Goffe's name and, as no money had yet changed hands, the sale was cancelled. Mr Edgar and his Chinese associates suffered considerable anguish over the incident.

Although the missionaries later withdrew their complaint, any possibility of securing the site was lost. The British Consulate-General continued to occupy a succession of cramped temporary premises for several years before finally leasing its own property in 1913. Even then, it seems to have been the embarrassment of enduring accommodation inferior to the French and German Consulates, rather than practical considerations, which prompted British officials to find themselves a permanent home in Chengdu.

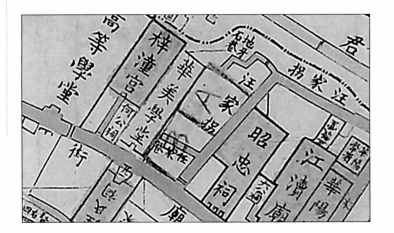

DESIRABLE NEIGHBOURHOOD: The site labelled A proved to be the cause of much discord between British diplomatic staff, Western missionaries and the Chinese authorities.

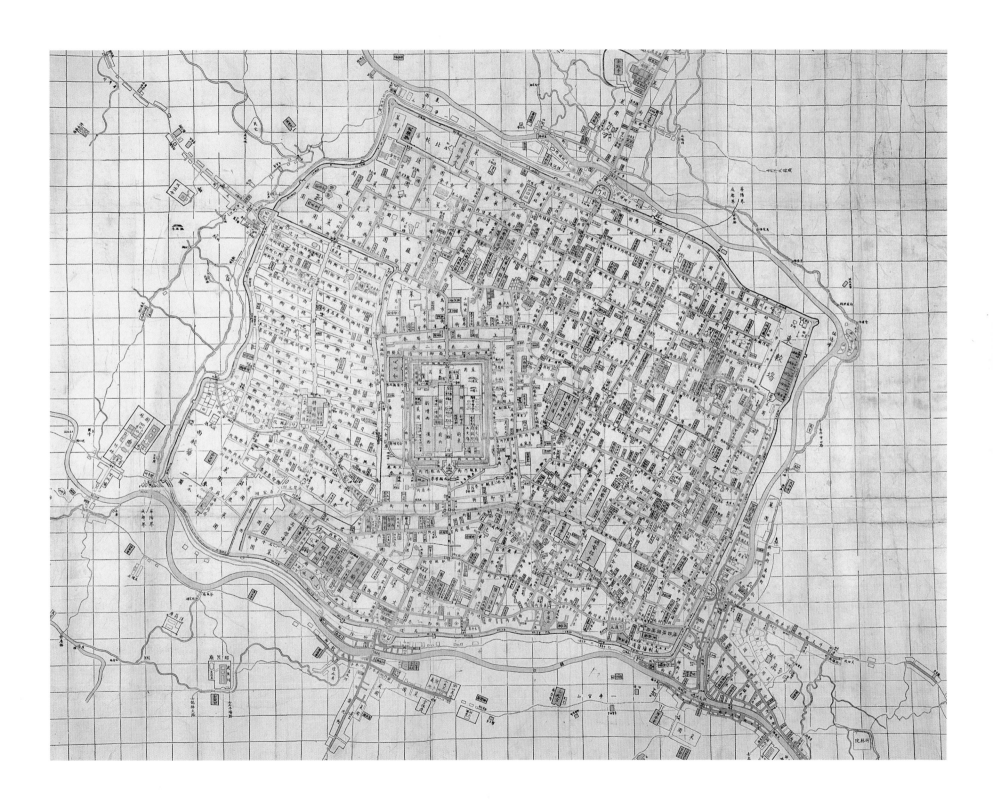

# Separate and unequal

Known as 'the Green City in the Sun', the cosmopolitan metropolis of Nairobi is now one of the most important urban centres in Africa. Yet little more than a century ago its site was just a swamp. The settlement originated in 1899 as a depot on the Uganda Railway, which was then being constructed to connect Lake Victoria with the Indian Ocean. The depot soon became the railway's headquarters and a township grew up around it. The slow transformation of Nairobi into a city began in 1905 when the capital of the British East Africa Protectorate (as Kenya was then called) was moved there from the port of Mombasa. This colourful map was created as part of the scheme devised by the British to expand the township.

Officials in Nairobi were deeply concerned about public health, particularly the risks presented by diseases such as malaria and bubonic plague. Professor William Simpson, of the London School of Tropical Medicine, was asked to advise the British on health and sanitary matters in East Africa in 1913. Like many medical experts of his time, he firmly believed that creating racially segregated residential districts within colonial towns was both natural and healthy. The early development of Nairobi reflected this – to modern minds very odd – assumption. Separate districts were planned to accommodate the city's three ethnic groups: the indigenous black Africans (many of whom were bitterly opposed to the creation of Nairobi), the white European colonists, and migrants from British India. The latter group consisted chiefly

of men who had originally come to East Africa to help build the railway and then settled in the colony with their families.

Papers sent from the Protectorate government to the Colonial Office in London in 1914 discuss how Professor Simpson's ideas were to be translated into reality. To illustrate this, coloured ink annotations were added to an outline plan reproduced by dyeline printing, a relatively cheap method of mechanical copying then quite commonly used for maps of this kind. According to these proposals, African residents would live to the south-east of the city centre (beyond the black outline) and Indians to the north-east (in the areas outlined in blue and brown). The largest area, to the west, would be reserved for the British community. Many of the green-edged protected areas were intended for European leisure activities, such as horse racing. Government-owned land shaded yellow could be sold or exchanged with other plots as necessary to encourage residents to follow the intended scheme.

The map captures the beginning of British policies in East Africa that were to create a three-tier society, in which European, Asian and African people formed the upper, middle and lower classes. Relations between the three ethnic groups were often tense, and the divisive effects of decades of inequality lasted long after Kenya became an independent country in 1963. Thankfully, enforced racial segregation is now no longer a fact of life in Nairobi. This map, however, remains a stark reminder of how colonial British ideas about urban planning were imposed onto an African landscape.

# Bombed out LONDON, 1940

The intensive bombing of European cities was one of the most devastating strategies deployed by both sides during the Second World War. Many old and beautiful buildings were completely destroyed; the survival of others, such as St Paul's Cathedral in London, seemed almost miraculous. Air raids caused many deaths and wreaked terrible damage upon the infrastructure of their targets, and thus successfully reduced the contribution that each city could make to its country's war effort. Nonetheless, the bombers failed in their second, and arguably more important, aim of crushing their opponents' morale. Neither the Germans nor the British were persuaded to surrender.

The London Blitz was among the most sustained of these bombing campaigns. The German Luftwaffe attacked the city more than 70 times between September 1940 and May 1941, dropping at least 18,000 tons of bombs and killing around 20,000 people. The map opposite is one of more than 650 prepared by the British Ministry of Home Security to record where these bombs fell. Such maps are just one aspect of the Bomb Census survey, conducted by the Ministry as part of its role to co-ordinate civil defence. Collating and analysing information about the patterns of air raids and their effect upon the urban environment – from individual damaged buildings to the broader impact on everyday life within the wartime economy – aided the authorities' efforts to administer the country in difficult circumstances.

In keeping with the 'make-do and mend' philosophy of the war years, staff at the Ministry used an array of official printed mapping when undertaking its survey. For the London Blitz, a special set of military maps, which employed a secret system of grid references to mislead the enemy, was pressed into service. Bomb sites were plotted onto these base maps by hand. Diagonal dashed lines symbolise showers of small incendiary bombs. Individual high-explosive devices are marked with dots; those thought to have been dropped in sequence from the same aeroplane are linked together. The circles with crosses represent the sites of unexploded bombs. This example covers part of central and east London during the week of 21–28 October 1940. The colours refer to different days of the week. These were not always used consistently: red, for instance, was normally assigned to Tuesday but signifies Saturday on this map.

The Bomb Census continued after the Blitz period. Initially confined to a few major cities, it was gradually extended across the whole of the United Kingdom as its value was recognised. The recordkeeping became more detailed as the war progressed and the government's thoughts began to turn towards post-war reconstruction. London's re-emergence from the rubble took many years. Many Londoners left the shattered central districts for the suburbs or in the new towns developed to house people migrating out of the capital. Although centralised planning was to alter much of its appearance over the following decades, one thing remained unchanged: London was still one of the world's greatest cities.

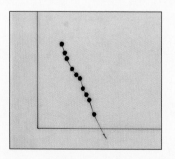

PLOTTING THE IMPACT: As well as annotating the printed maps, Ministry staff prepared tracing paper overlays showing the bombs that fell on specific days. This tracing refers to the 24-hour period from 6 am on 26 October 1940.

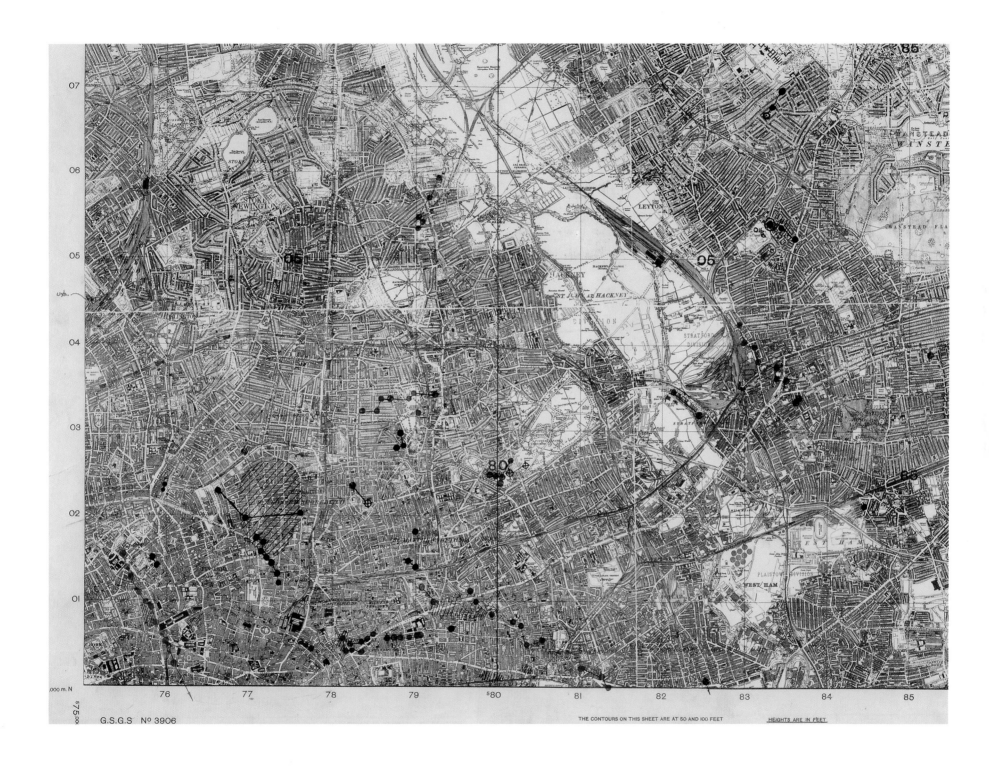

# CHAPTER THREE

# THE COUNTRYSIDE:
# landscapes
# in time

If the landscape is its own map for those who know how to read it, maps themselves offer guided views of the landscapes they portray. They allow us to survey larger areas such as counties and provinces, and to appreciate the fine detail of change and continuity in localities. They give clues to help unlock the mysteries of the countryside by providing evidence that features existed and what they looked like at a point in time, and they help us to picture how people lived and worked in past places. Where new towns have arisen and cities expanded over the centuries, earlier maps serve as a record of fields and woods before they were lost under a sea of brick and stone.

Maps of the countryside show what the mapmaker was commissioned to include. The 17th and 18th centuries were the golden age of manuscript estate maps, which are often highly decorative symbols of status and power. A new class of professional surveyors were employed by a single landowner or by bodies such as charities, colleges, companies or the Crown. Lands often changed hands in these centuries of political upheaval, when backing the wrong cause could lead to loss of one's estate as well as one's head, as we see in a Northumberland case (page 87). The map of Audley End on page 85 was made to celebrate Charles II's acquisition of that quasi-palace. These large-scale estate maps, with accompanying written surveys which detail fields, tenants, rents, and crops grown, together informed good estate management. But these maps offer only a partial view of a piece of countryside. Names of adjacent landowners may be shown, but not their lands. Only more attractive features were included, not dunghills and eyesores.

Britain's expanding empire overseas opened up new lands and new landscapes to be surveyed. The estate map was transported and transformed as a land-grant plat or a plantation plan of new-cut colonial lands. The style and layout of these maps were similar, whether in Georgia or South America, as we can see from the maps on pages

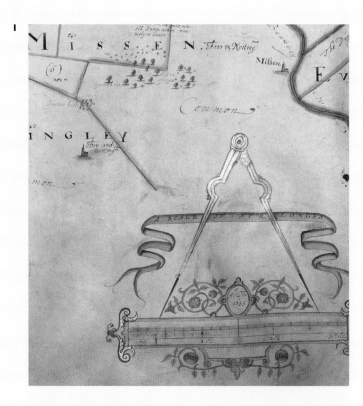

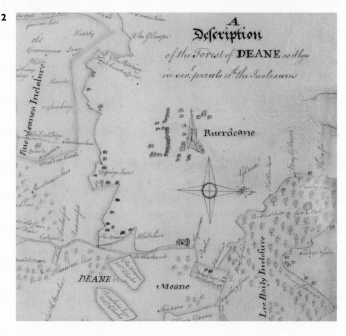

**1** DECORATIVE DIVIDERS: Estate maps
are drawn at a large scale.

**2** ON THE FOREST FRINGE: This map
record the forest edge, an area of transition.

**3** CHINA: THE DISTANT MOUNTAINS:
View of a river city in Guangxi Province
about 1850.

90 and 88. However, they record a more exotic type of native trees and plants, with crops such as sugar, coffee, indigo and rice. Whether at home or abroad, estate maps tell a specific story about the ownership, exclusivity and productivity of a limited area, reflecting a social structure based on the land. Other maps record man-made changes in the landscape such as drainage schemes (the map on page 83), the making of canals and railways, or modernisation by enclosure to make large fields instead of small medieval farming strips.

From the later 18th century there was a move towards detailed maps of the wider landscape. This trend is seen in a county map of Sussex (on page 93), a military survey of Guadeloupe (page 95) and a hunting map of part of Spain (page 101). These kinds of topographical map offer more comprehensive cover of the countryside, including wastes, moorland and marshes of little or no agricultural use, and so they fill in the gaps between estate maps. Tithe Survey maps were made for many parishes in England and Wales in the mid 19th century (see the Cornish example on page 99). The demise of the purpose-made estate plan came with the wide availability of printed large-scale maps, which could be customised by means of manuscript additions. The archives holds two later national surveys taken after the Tithe Survey; the Valuation Office Survey of around 1910 and the National Farm Survey made in the Second World War, which both used the latest Ordnance Survey maps, marked with hand-drawn survey detail.

The fascination of these maps of estates and the wider land lies in their portrayal of changes in the rural landscape, and the way that they show features from different eras on one sheet. Ancient footpaths and drove-roads connect abandoned settlements, alongside long-empty canals and turnpike roads. Old stones on remote moors stand within sight of churchyards. Early open-cast mines and quarries rest alongside iron foundries and factories of the Industrial Revolution. The sea overwhelms or recedes from the land, leaving submerged villages or dry sea ports. Whether village green or market square, forest or river course, where there are successive maps we can gauge the pace of change in these places. These maps, along with related textual records, contribute to our understanding of how countryside evolved, and allow us in imagination to step into the landscapes of times past.

**4**

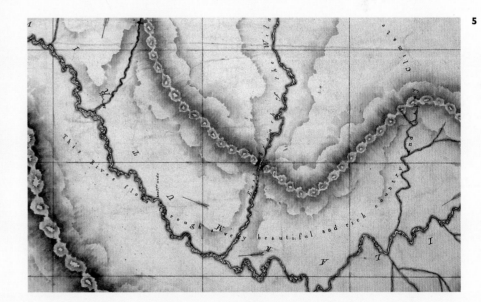

**4** UNCHANGING LANDSCAPE: An estate map of 1775 was reused as a tithe map in 1839. The field pattern did not change, just the landowners.

**5** RIVER DEEP, MOUNTAIN HIGH: One of the explorer David Thompson's maps of north-west Canada, which notes 'the happy climate'.

**6** VISTA ACROSS THE RIVER TAW: This watercolour view was painted on a map of a Crown estate at West Ashford in north Devon in 1812.

**5**

**6**

# 'All such necessarie things'   ALDBOURNE CHASE, WILTSHIRE, 1608

This is the central part of a map of Aldbourne Chase which lies northeast of Salisbury Plain among the chalklands of the Marlborough Downs, in White Horse country. The map was drawn in a dispute about the royal rabbit warren which lay within the Chase. Words cleverly concealed in banks of trees across the map state 'Southe Ood Walk' which refers to Southwood Walk, one of three areas into which the warren was divided. Although this map was made for a legal case, we know from the title that it aimed to show a wider picture, including 'the landes neare adioyninge' as well as the contested area. The instructions for making the map required it to show 'all such necessarie things as you … shall thinke beste for the more manifestacion of the truthe of the said matter'.

The lordship of the manor of Aldbourne was held by the Duchy of Lancaster. Its Council ordered the map to be made as part of an investigation into a dispute centred on land management, boundaries, common land and access rights. The Duchy alleged intrusion on its land by George and Roger Walrond, members of a family who were hereditary Keepers of the Chase. At certain times of year the Walrond brothers enclosed a field (at lower left) to grow corn, claiming that the terms under which they held their farm allowed them to do this. The Duchy's tenants complained that this effectively barred them from grazing their animals there.

The map and court papers reveal a complex pattern of ownership and land tenure. Named fields are labelled with their owner or occupier, whether royal land, the Walronds' inheritance or copyhold of the manor. Their use is given, as arable, pasture, coppice, woods or closes. Each piece of land also has a pictorial representation of the flora and fauna found there, crammed in a mass of detail over a plan drawn to scale. Just as larger and smaller human interests were interwoven in any one parcel of land, so the animals interact with the trees and bushes growing there. The mapmaker has shown cattle, sheep and horses grazing, some with their heads hidden behind the herbage.

The mapmaker 'thought best' to draw a hugely detailed picture of this swathe of countryside. He records a lodge built in 1606, and the remnants of a medieval ridge-and-furrow farming pattern, although these were not relevant to either party's case. Notes outside the map explain that the green lines denote enclosures with quickset hedges, while dotted lines indicate common land. Each tree is shown with its canopy in the shape of its leaf; oak and ash are immediately obvious, and others may be elder, hazel, hawthorn and willow. There are red-flowered plants, scrub, underwood, even dying trees, in a scene which certainly brings us 'the more manifestacion' of this place, one summer over 400 years ago.

FLOWER BORDER The court instructed that the map was to be 'framed'. This section shows the decorative running pattern of pinks – Shakespeare's 'streak'd gillyvors' – intertwined with peas in their pods, fruiting grape vines, red flower buds (perhaps sweet peas?) and acorns, echoing the oak trees drawn on the map.

# The dangers of decoration THE FENS, EAST ANGLIA, 1658

The changing tides of history have brought only one example of a large and impressive 17th-century engraved map safely to the shores of later times. This is one sheet of 16 which were made to celebrate the achievements of an ambitious seven-year engineering project to drain the Fenlands and so turn waterlogged flood-prone land into good farming soil. The National Archives holds the only known copy of the first edition. Why did more copies not survive? The timing of its production, and its distinctive decoration, together told against it.

Around the edge of this map range 87 coats of arms of investors – 'adventurers' – in a private venture company set up to drain the Fens during the 1650s, when England was ruled by Oliver Cromwell and the Roundheads. Many of these arms-bearers were Parliamentarians and military officers who had served in the New Model Army during the Civil War. The two coats of arms shown are those of the regicide William Goffe and Sir Walter St John, son-in-law of Cromwell's Lord Chief Justice. The decoration thus gave this map a political complexion.

This sheet demonstrates how the draining of the Fens physically changed the face of the landscape and the flow of water across it, in accord with the mathematical principles dear to Jonas Moore, the scheme's surveyor and mapmaker. Straight lines provided direct means for water to run, replacing meandering channels. From the top edge where lay The Wash, Downham Eau was cut straight, alongside the winding 'Old Ouse'. Below a complex junction of waterways at Denver lie the parallel Old and New Bedford Rivers, with raised banks to contain floodwaters between them. To left the 'Marshland Cutt' drove straight across the land. Below it is the 'new-bottomed' Popham's Eau. New sluices, channels and embankments are evident all over this map. The works also changed ownership; this sheet shows some of the new 'lots' given to investors, created from newly-drained lands, among existing fields and commons.

This expensive map would have adorned the walls of the rich and investors in the scheme – precisely the people who needed to prove their loyalty to the King when the monarchy was restored in 1660, two years after the map's publication. A map boldly displaying coats of arms of known Parliamentarians would hardly help their cause. Presumably all copies were destroyed except this one, sent to the Duchy of Lancaster, the area's major landowner.

The coats of arms were erased from the printing plates, just as the Commonwealth period was superseded by the Restoration, and the drainage scheme itself had limited long-term success. Yet the quality of the map launched the career of Jonas Moore, who became Surveyor General of the Royal Ordnance, a leading light of the Royal Society, friend of Sir Christopher Wren, and a founder of the Royal Observatory at Greenwich. The map itself was a landmark in large-scale surveys, reprinted with little change for another 150 years – minus those contentious coats of arms.

NOW YOU SEE IT … The same area of a later version shows where the coat of arms had been removed from the printing plate.

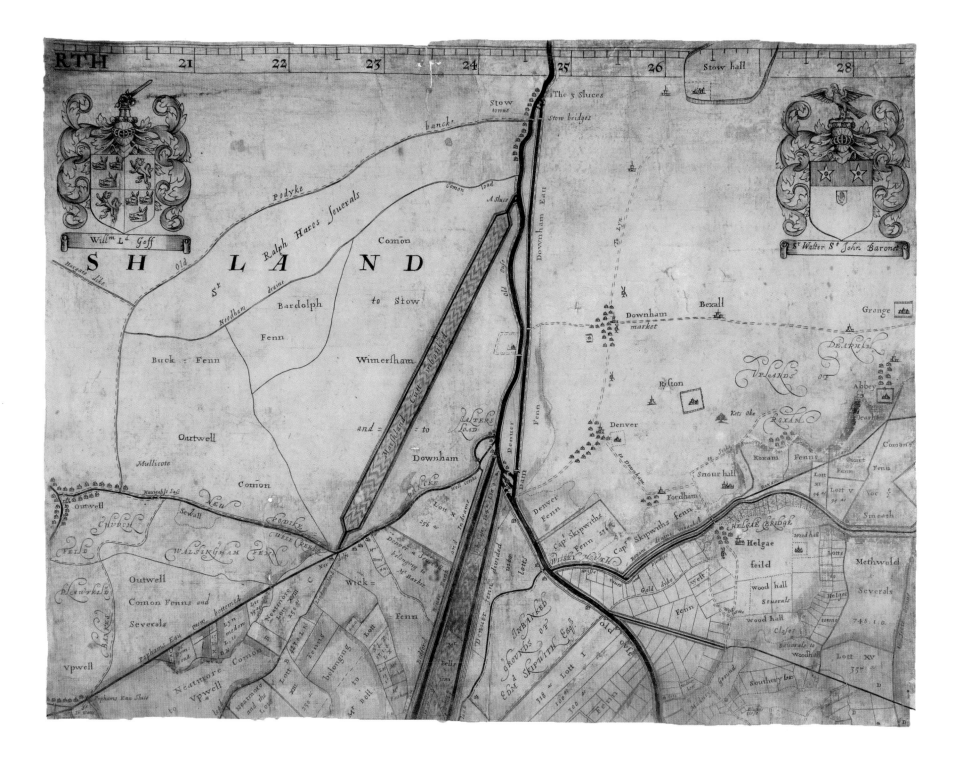

Will.ᵐ Lᵈ Goff

Sᵗ Walter Sᵗ John. Baronet

SH    L A    ND

Podyke

Ralph Hares Seuerals

Sᵗ

old

Hargate dike

Needham draine

Bardolph

Buck = Fenn

Fenn

Comon

to Stow

Wimersham

and = = to

Downham

Oartwell

Mullicote

Comon

Nauigable Salt

NEW

Outwell

Sewall

PODYKE

CHURCH

FEILD

CHUL CREEK

WALSINGHAM FENN

Naye delfe

BEAWFELD

Outwell

Comon Fenns and

WICK =

Fenn

Severals

Vpwell

Pophams Eau Sluse

Neatmore Comon

Vpwell

A Sluce

Comon road

A Sluce

Downham Eau

old Pode

Marshland Cutt intended

Denver

Dam

Fenn

Salters Lode

Downham

BECKE

Lott X

256 ac.

Denver a Dam
belonging to
Mr. Bacher

Denver fenn diuided
in
lotts

INBANKED
GROUNDS OF
Edm Skipwith Esq.

Lott I

Denver
Fenn

Cap.ᵗ Skipwiths

Wisser Fenn 256 ac.

Cap.ᵗ Skipwiths fenn

WISSEY MOUTH

old owse

Gold dyke

Stow
towne

The 5 Sluces

Stow bridges

Downham
market

Bexall

Ryston

Kets Oke

Denver

Snour hall

Fordham

Grange

DEARNH

VPLANDS OF

ROXAM

Dearnhill

Abbey

Comon

Court
Fenn

Fenn

Roxam Fenns

Lott
24 ac

Lott V
39 ac

Voc
5

Sineoth

HELGAE BRIDGE

Helgae
feild

wood hall

Severals

wood hall

Closes

Wood hall

Lions

Methwold

Helgai

Severals

748.1.0.

Southery low

Lott XV

357.

D

Woodhall

D

# Across four centuries; triumphs and disasters

The history of the house shown centre right on this attractive map, and that of the accompanying photograph of the map are marked by parallel tales of schemes thwarted. The house was planned to forward a personal ambition, while the photograph was part of an enterprise designed to safeguard the nation's heritage. The map links these two stories.

It is 400 years since Audley End House was built by a rising courtier who sought to impress. Its reputed cost was £200,000 – about twenty million pounds in today's money. James I, when he saw it on completion in 1614, reportedly commented that the building was 'too large for a King, though it might do for a Lord Treasurer'. The Earl of Suffolk, its builder, was in fact James's Lord Treasurer, and soon was imprisoned for suspected embezzlement from the royal coffers to help finance the building. This ended his grand plans for his family's advancement.

This map was made in 1666, when Charles II decided to buy the palatial house, despite his grandfather's remark, to stay in when he visited nearby Newmarket racecourse. The map states that it is by George Sargeant, then well known for mapping private and Crown lands, but accounts for work on royal buildings show that in May 1666 11 pounds was paid to one Maurice Emmet and an assistant for 16 days preliminary measuring and survey at Audley End.

The newly royal property is presented with panache on this typical 17th-century estate map, the house and park set within a number of colourful decorative elements. A border of leaves, flowers and berries runs around the outside. The large compass rose points to north at the right of the map, with the town of Saffron Walden to the west, shown at the top of the map. Ornate fretwork cartouches enclose the map's title, a table of acreage for

areas of the estate, and a scale bar at lower left. The grounds mix ornamental and productive areas, with cherry and rose gardens, 'milk yard', brewhouse, barns, a dovecote, hop ground, and even a vineyard and bowling alley. The park beyond reflects a trend to formality, with avenues of trees.

The map was photographed in January 1933 at the Ordnance Survey office in Southampton, as part of a scheme to capture images of important maps. The Archaeological Officer whose stamp appears on the photograph was OGS Crawford. He had served in the First World War and, aware of how much destruction of cultural property it had caused, was keen to create a centrally-held photographic record, in case the original maps were later lost. By an ironic twist of fate, it was the records of Crawford's scheme that were themselves destroyed by bombs in the Second World War. If the ambitions of Crawford and of the Earl of Suffolk were not realised in the way that they had hoped, then the house, this print of the photograph, and this splendid example of an estate map survive today, as testament to their endeavours.

SAFETY SHOT: This photographic print was presumably sent with the original map when it was returned to the Commissioners of Crown Lands, after it had been borrowed by OGS Crawford to have photographed at the Ordnance Survey office. The scale bar shows it was made at a reduced scale to the map.

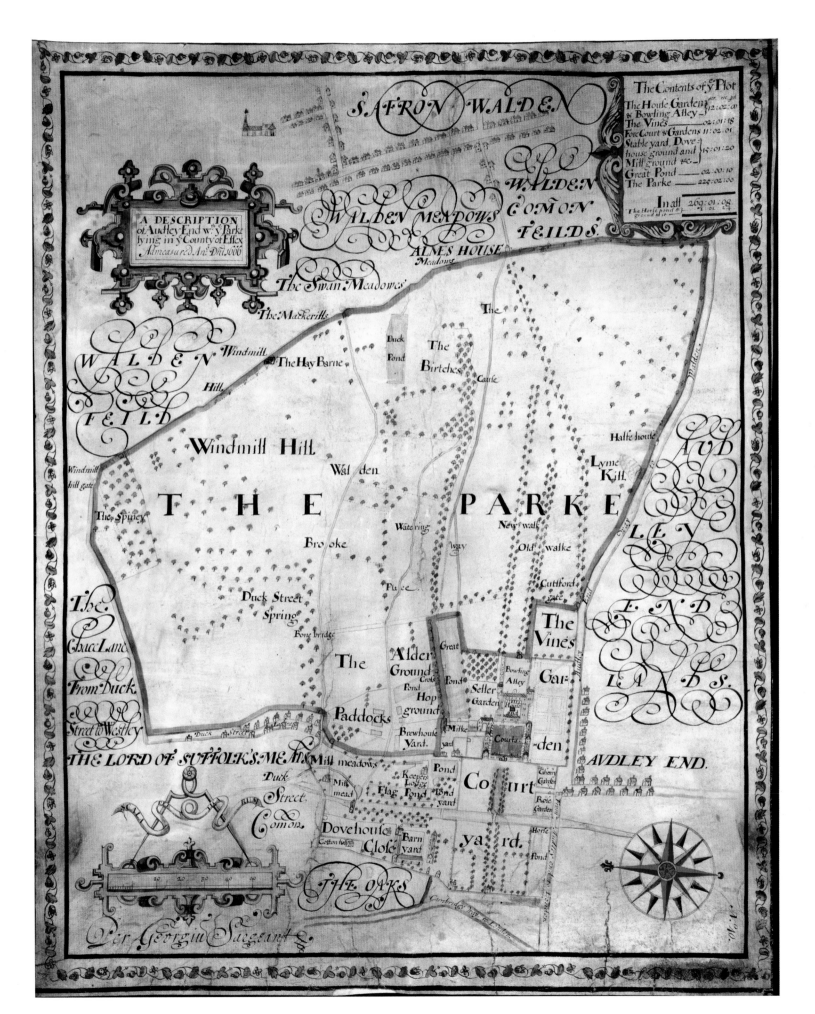

A DESCRIPTION
of Audley End w.th y.e Parke
lying in y.e County of Essex
Admeasured An.o Dn.i 1666

S. AFRON WALDEN

WALDEN MEADOWS

WALDEN COMON FEILDS

The Contents of y.e Plot
The House Garden & Bowling Alley — 12:02:0
The Vines — 02:01:18
Fore Court & Gardens 11:02:01
Stable yard, Dove-house ground and Mill ground &c. — 15:01:20
Great Pond — 02:00:10
The Parke — 225:02:00

In all 269:01:08
The Horse pond & ground to it — 01:01

ALMES HOUSE Meadow

The Swan Meadowes

The Mackerills

WALDEN FEILD

Windmill

Hill

The Hay Barne

Windmill hill gate

The Spiney

Windmill Hill

Walden

THE

Brooke

Duck Street Spring

Pong bridge

The Chace Land

From Duck.

Street to Westley

THE LORD OF SUFFOLK'S MEAD

Duck Street Comon.

Per Georgiu Sargeant

Duck Pond

The Birtches

Cause

Halfe houfe

Lyme Kill.

PARKE

Watering

Way

Place

New walk

Old walke

Cuttford gate

AUDLEY END LANDS

The Vines Gar-den

Great Pond

Crofs Pond

Hop ground

Brewhoufe Yard.

Setter Garden

Bowling Alley

Milke yard

Courts

AUDLEY END.

The Paddocks

Alder Ground

The

Mill meadows

Mill mead

Keepers Lodge

Flag Pond

Pond yard

Pond

Co urt ya rd.

Cherry Garden

Rose Garden

Horse Pond

Dovehoufe Clofe

Barn yard

THE OAKS

# A confiscated Jacobite estate

Why is a set of estate maps of northern England found among Admiralty records? The lands in question were those of the Third Earl of Derwentwater, whose mother was an illegitimate daughter of Charles II. The Earl grew up at the Stuart court in exile in France as companion to the young prince James, and his part in the attempt to restore James to the throne in 1715 by force of arms led to his execution for treason by the Protestant George II. The Earl's lands in Cumberland and Northumberland were seized by the Crown and given to support Greenwich Hospital for old Royal Navy men, which was administered by the Admiralty.

The Hospital had a survey made in 1736 to record the extent and value of their acquisition. This is one of 34 manuscript maps which illustrate the survey, showing land in a number of parishes. These were not just agricultural estates; there were slate and limestone quarries, collieries, mines and mills. The surveyor Isaac Thompson states in a frontispiece to the volume of maps 'the method used to express ye most remarkable Things in ye Survey'. He gave a letter to each farm so all its fields could be easily distinguished, and colours show where distinct fields adjoined. All the main features 'are copied from ye Things themselves'. A key to symbols includes those for stiles and gates, springs and wells, carriageways and footpaths, boundary stones, and traces of 'destroy'd' hedges and fences.

Rather than showing all the land in a parish or specific area, as is the case with the maps seen so far, these are classic examples of estate maps. They only show the property of one landowner, whose parcels of land may be scattered across parishes and even counties. They indicate who the surrounding landowners were, but supply no detail of those lands. On other sheets of this survey isolated fields appear as islands in a sea of blank parchment; this is a very selective view of the landscape. The maps were for use with the written survey, which gives for each field the tenant's name, the field name and acreage, and its use as arable, meadow, pasture or woodland.

Woodhall is a small place near Hexham, south of the River Tyne shown on the map, a few miles south of Hadrian's Wall. The map's focus is the land, within a framework of the river and roads, bordered by lands of adjoining estates. Running through the lower centre of the map is a clough, a narrow wooded ravine typical of this area. Very small pen and ink detail shows houses, mills and a quarry. The fields labelled either 'H' or 'M' belonged respectively to Woodhall Farm or Woodhall Mill. Red ink additions to Thompson's maps note later changes such as gains to the estate by gift or purchase, land divisions, and the effects of Enclosure Acts in the 1790s. This is typical of the way that an estate map often continued to serve across time.

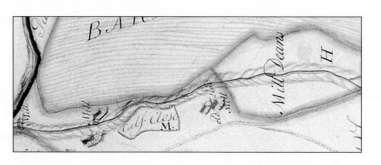

VALE OF INDUSTRY: Lead mining was a major industry in this area. Lead would be turned into ingots in smelting mills along waterways, as here.

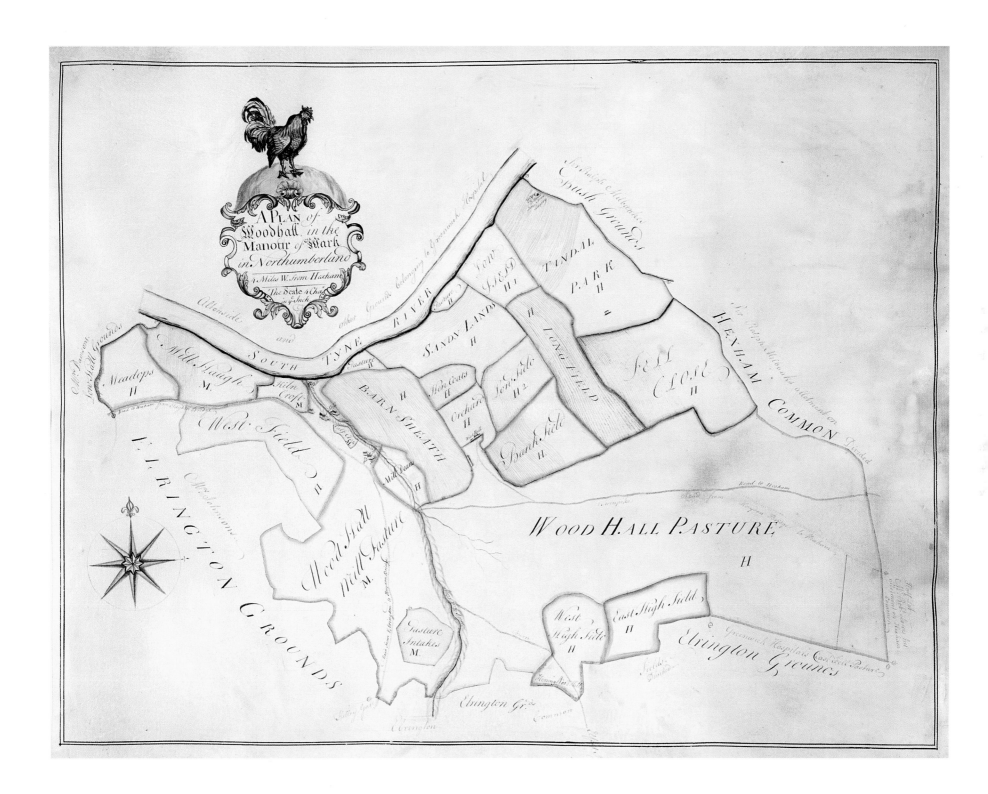

A PLAN of
Woodhall, in the
Manour of Wark
in Northumberland

4 Miles W. from Hexham

The Scale 4 Chains
to ½ Inch

# Going Dutch

PLEGT ANKKER, BERBICE, GUIANA, 1768

In the late 18th century there was much exchange of colonies between European powers, and it was usual practice for the departing nation to take away its records. The Dutch West India Company had operated in parts of South America since 1621 and established trading and sugar-farming colonies in Demerara, Essequibo and Berbice, all on the continent's north coast in the area which became British Guiana and is now Guyana. When the British took over this region in 1815 they asked for the transfer of official records along with the lands, to help the new administration, and the Dutch agreed. The papers left behind include court records, taxation returns, reports on slaves, and grants of land.

Among these Dutch archives was this plan, bound in a volume of similar items, each accompanied by a dated certificate granting the land shown. These are mostly plans of small estates. In a dedication at the front of the volume, the colony's Surveyor General observed that he was thus delivering copies of the measured surveys to the Governor as promised. This explains why the plans are homogenous in style although their stated dates range from 1756 to 1771. There are also two miscellaneous plans which would both have interested the Governor. One shows a fort where gunpowder was kept, and the other a building which may have been the government house. This volume as a whole offers an insight into the work and abilities of an official surveyor appointed to oversee land survey in an established colony.

This plan shows the 2000-acre plantation of Plegt Ankker, edged in gold at the top of the sheet. The straight boundaries are consistent with a colonial land grant. This property of Jan de Koning lay upriver from New Amsterdam and south-east of Georgetown, on the River Berbice. Two of its smaller neighbours have been included below it, perhaps to give a better idea of where Koning's land lay in relation to the bend in the river. Along the left-hand boundary the colony's own ground is named but not shown in detail. The Dutch had a long tradition as seafarers and chart-makers, which may explain the presence of quite a large circular compass indicator, strategically placed so that its centre is where the southern boundary of Plegt Ankker met the north-west corner of the estate edged in blue, Christina's Lust.

The scale is given in Rhineland roods, the most common Dutch land measure, where one rood was equal to just over 14 square metres. Notes written across the map indicate that Plegt Ankker was not an ideal plantation. There was some higher better ground and woods, but mostly the soil was waterlogged clay and peat in a marshy landscape. The surveyor captured all these plantations in several volumes representing the colony's land. Meanwhile, in a fanciful touch, the local river god – who as fount of the stream symbolically sits leaning on an overflowing ewer – waves an oar as he surveys his watery domain.

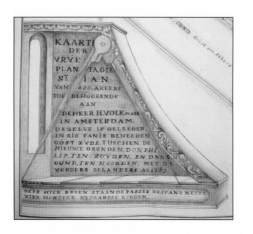

THE MEASURE OF THE LAND: It is likely that the surveyor added decorative touches such as this when he copied original plain plans, for the Governor's view.

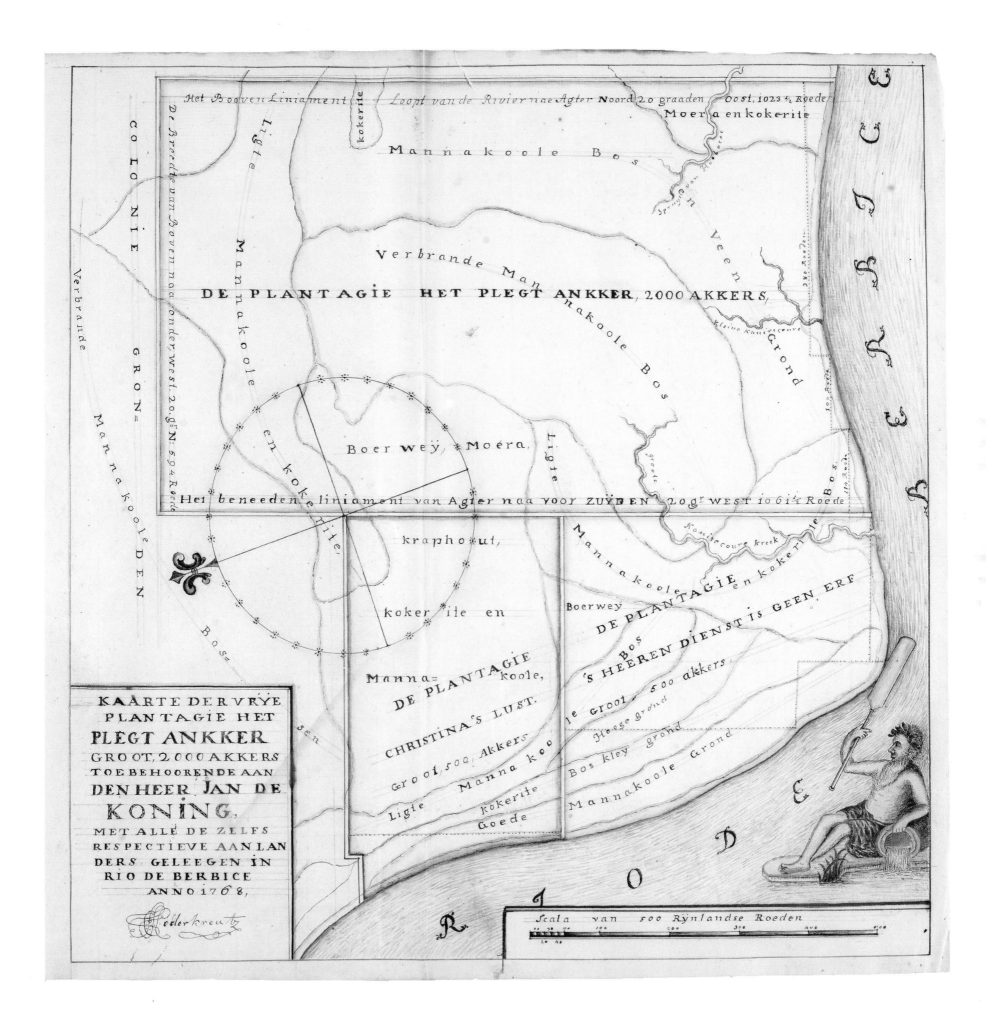

BERBICE

Het Booven Liniament Loopt van de Rivier nae Agter Noord 20 graaden Oost, 1023 ½ Roede

Moera en kokerite

CO LO NIE GRON=

De Breedte van Boven naa onder West 20 gn Ns 594 Roede

Ligte Mannakoole en kokerite

Verbrande Mannakoole Den

Mannakoole Bos

Verbrande Mannakoole Bos

DE PLANTAGIE HET PLEGT ANKKER, 2000 AKKERS,

Veen

Grond

Boer wey, Moera,

Ligte

kleine Kanjecoure

groote

Het beneeden liniament van Agter naa voor ZUYDEN 20 gr WEST 1061 ½ Roede

Kanjecoure kreek

kraphout,

Mannakoole en kokerite Bos

koker ite en

Boerwey

DE PLANTAGIE

'S HEEREN DIENST IS GEEN ERF

Manna= koole,

DE PLANTAGIE

Bos

Mannakoole Groot 500 akkers

CHRISTINA'S LUST.

Hooge grond

Groot 500 Akkers

Bos kley grond

Ligte Manna koole

Ligte Kokerite Goede

Mannakoole Grond

RIO DE

KAARTE DER VRYE
PLANTAGIE HET
PLEGT ANKKER
GROOT, 2000 AKKERS
TOEBEHOORENDE AAN
DEN HEER JAN DE
KONING,
MET ALLE DE ZELFS
RESPECTIEVE AANLAN
DERS GELEEGEN IN
RIO DE BERBICE
ANNO 1768,

Scala van 500 Rynlandse Roeden

# A swamp in time

While the Northumberland estate map (on page 87) was drawn to record lands gained, this map records lands lost by an English colonist in America after the War of Independence in 1783. The British Loyalists' Claims Commission was set up to enquire into such property losses and to decide which claims merited compensation. The map was part of a claim for £33,000 by Samuel Douglas for a number of properties and was filed with other maps and papers submitted as evidence to the Commission. Douglas was among those eventually rewarded for his loyalty and his losses, if not to the extent he claimed.

The map conveys a lost era in the sense that the place-names and the world they connote of English colonial planters were both banished not long after it was made. Demere's Island, named for Captain Raymond Demere who was given the original grant in 1760, was renamed Champney island. It lies in the delta of the lower River Altamaha (Alatamaha on the map) near the Atlantic coast. North of Demere's Island, along the right-hand edge, lay the islands of Lachlan McIntosh and Governor Wright, now Rhetts and Butler Islands respectively. Perhaps these places were the inspiration for Georgia-born author Margaret Mitchell's leading man in *Gone with the Wind*? We do know that James Wright was Georgia's last governor, and that he also submitted a large claim. One name that has remained, despite being a tribute to the English King George II, is that of Georgia itself.

Demere's island, under whatever name, is as the map shows, enclosed in two branches of the river, which is tidal at this point. 'Cane swamp' may refer to sugar, which was grown on this coast, while the oyster banks point to another local industry. The marshes would have been of little agricultural use, although they support wildfowl. The island is shown with trees; those mentioned on maps of nearby estates show hickory, oak, maple, black walnut, gum and pine. These are confirmed by the diaries of the naturalist William Bartram who recorded the flora and fauna when he paddled these parts of the Altahama River in a cypress canoe in 1773. Along with alligators, he found a rare flowering tree which he named *Franklinia alatamaha* and which survives only in cultivation as plants descended from seed he collected.

The text on the map shows that it was certified in 1771 as a re-survey, which clarified that the island contained 1,000 acres: the original grant for 560 acres to Demere, plus a further 440 acres to John Graham, who was Lieutenant Governor. The text notes that the island lay in St Andrew's, one of Georgia's eight parishes. As in England, parishes supported their church, and also served as districts for elections, taxation, welfare and other tasks of the secular state. If the administrative framework was similar, this example suggests how the estate map too was transplanted to the colonies, to depict different landscapes and different ways of cultivating the land.

FACES IN THE SWAMP This is one of a number of similar plans which all bear these whimsical compass indicators. Each face is different; perhaps this portrays the mapmaker or someone he knew?

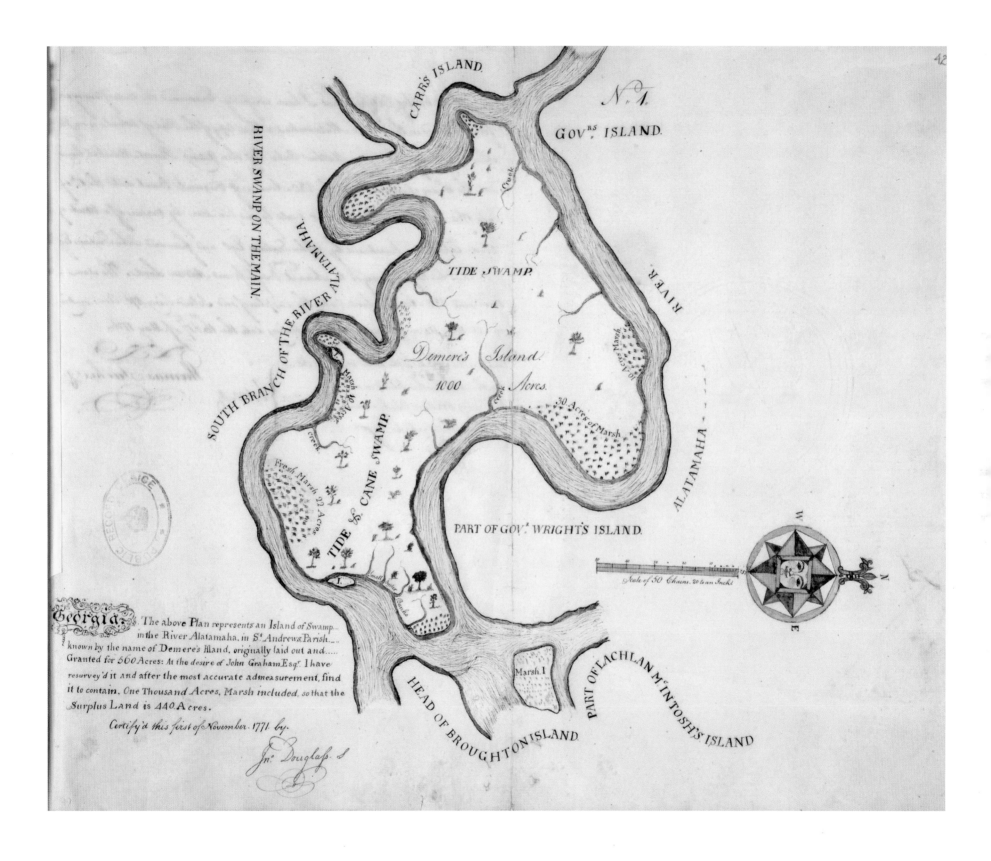

CARR'S ISLAND.

GOVRS ISLAND.

N.º 1.

RIVER SWAMP ON THE MAIN.

ALATAMAHA

SOUTH BRANCH OF THE RIVER

RIVER

TIDE SWAMP.

Creek

Demere's Island.
1000 Acres.

30 Acres of Marsh

10 Acres Marsh.

creek

Marsh 10 Acres

creek

Fresh Marsh 22 Acres.

TIDE & CANE SWAMP.

Small Salter Creeks

PART OF GOVR WRIGHT'S ISLAND.

ALATAMAHA

Georgia. The above Plan represents an Island of Swamp
in the River Alatamaha, in St Andrews Parish
known by the name of Demere's Island, originally laid out and
Granted for 560 Acres: At the desire of John Graham Esqr. I have
resurvey'd it and after the most accurate admeasurement, find
it to contain, One Thousand Acres, Marsh included, so that the
Surplus Land is 440 Acres.

Certify'd this first of November. 1771. by.

Jno. Douglass S.

Scale of 50 Chains. 20 to an Inch.

Marsh I.

HEAD OF BROUGHTON ISLAND.

PART OF LACHLAN MCINTOSH'S ISLAND.

# Towards a national survey

Today we are used to the fact that Ordnance Survey provides mapping for the nation. This was not always the case, and this map marked an important step towards the kind of accurate, scientific maps that we now expect. It was an early product of an important partnership in British cartographic history. The 3rd Duke of Richmond employed two young men, Thomas Yeakell and William Gardner, as surveyors of his large estate at Goodwood, where the lattice of the park is shown above centre. Through the 1760s and 1770s they mastered the art of survey by mathematical methods here, and then sought a wider canvas beyond the estate walls on which to practice their art.

The first result was this engraved map of Sussex at a scale of 2 inches to 1 mile. This is the westernmost sheet of four made, out of the eight originally proposed. Ranging from the Hampshire border at left across to Arundel, it shows the network of fields and woods, with Chichester coloured red as the only town of any size. The South Downs across the top are portrayed in an eye-catching way by dark hachures which emphasise their height. The coast is shown incised by harbours and inlets, with good anchorages noted. Both men carried out the survey work, and Yeakell also engraved the map. This county map project faltered because the maps were expensive and did not sell well. Its surveyors, too, were drawn into work elsewhere, on the national scale that they desired.

The Duke of Richmond was Master of the Board of Ordnance in the 1780s and 1790s, responsible for all aspects of supply to the army, including maps. It was from this Board that Ordnance Survey took its name. As a military man who owned an estate on the south coast and valued maps, Richmond proposed a national topographic survey. This would offer defence information about the lie of the land and good landing places, should France – then heading for Revolution – try to invade England. While arguing for such a survey at the highest levels, the Duke ensured that his protégés, Yeakell and Gardner, assumed senior positions among the Draughtsmen in the Tower of London, where the Ordnance office was based.

The Board's records include a rare survival of 'foul drawings', draft manuscript maps made in the field, of parts of Sussex and Kent by Yeakell, which throw light on the mapmaking process. Doodles show where he tried out his pen, did calculations and practiced writing words, sometimes unrelated to the map: 'Tuesday', 'walnut' and 'dang' appear. From such sketches arose finished hand-drawn coloured fair plans, on which the engraved maps were based. Gardner outlived Yeakell and went on to supervise production of the first Ordnance Survey maps in the late 1790s, drawing on the partners' earlier work on this map but using new survey instruments and methods of triangulation. Thus, this map and its creators contributed to the genesis of a national survey.

FOUL DRAWING Part of a rough sketch map which shows the first stage towards the finished map. Detail of hills and fields is accompanied in the margin by drawings of a church in both plan and elevation.

THORNEY ISLAND

Pilsey Sand.

BRACKLESHAM BAY

THE PARK
Good Anchoring Ground

PAGHAM HARBOUR

BOGNOR ROCKS

Midleton Ledge

Shelly Rock

Middle Ledge

B.O

SCALE of Three Statute Miles.

3m 16s. of Time West of the Royal Observatory at Greenwich.

# 'Low Land' BASSE-TERRE, GUADELOUPE, 1793

By the end of the 18th century, Britain's military leaders were learning to value systematic knowledge of any ground on which action might take place. Just as this was important at home for defence of the English Channel coast, so it informed her overseas campaigns. At the same time, her allies and enemies alike were becoming aware of the same lesson. The French were cartographically advanced; they already had a printed national survey and their military mapmakers were producing work of exceptional quality, especially with regard to height depiction. These standards transported to their lands overseas.

Between 1793 and 1815 Britain and France were at war for most of the time. The conflict in Europe was mirrored in their colonies, especially in the Caribbean where a number of islands changed hands, some more than once. The archives thus holds maps such as this one of islands usually considered part of the French empire. Guadeloupe had been colonised by the French in 1635, and is still an overseas region of France today. This manuscript map was made by the French in preparation for a British attack which came in April 1794, when the map was captured along with the island. France retook Guadeloupe the next year for good, except for a further British occupation from 1810 to 1815. A map such as this which shows a theatre of conflict in great detail was useful for defender and invader alike.

The map's title states that it is a topographic plan of part of Guadeloupe between the Rivers of Baillif (lower left) and of Grande-Ance (top right). The town of Basse-Terre is in red in the centre of the coast, with Fort St Charles above it (to its right), briefly renamed Fort Matilda by the British between April and December 1794. The map was annotated by the British military to show the forces of named generals. These include Thomas Dundas who was briefly the island's governor but died there of yellow fever in June 1794.

Beyond its narrower military use the map is a hugely detailed topographical study. It shows part of the western of Guadeloupe's main islands, called Basse-Terre, or 'low land'. Despite its name, which refers to its situation leeward of the trade winds, this island is very mountainous. At right is the darkly-shaded La Soufrière volcano at 1,467 metres above sea-level, higher than Ben Nevis, and the tallest mountain in the Lesser Antilles. The map shows how the dominating rugged relief is interspersed with forests and grasslands, with fertile flatter lands along river valleys and the coast. It was here that the island's sugar plantations lay.

This map travelled to London and became part of the Colonial Office map library, filed at a point when Guadeloupe was British for a short time. It joined maps from the other brief British invasions of the island. Whatever the original reason it reached the archives, the detailed topography and exquisite draughtsmanship continue to command attention.

**ENCOUNTER WITH NATURE:** This sketch shows a British officer, sketchbook in hand, recording the landscape and lush vegetation around Nosier Ravine. This was probably Lieutenant Colonel Light, who drew a series of views during his service in Guadeloupe in about 1815.

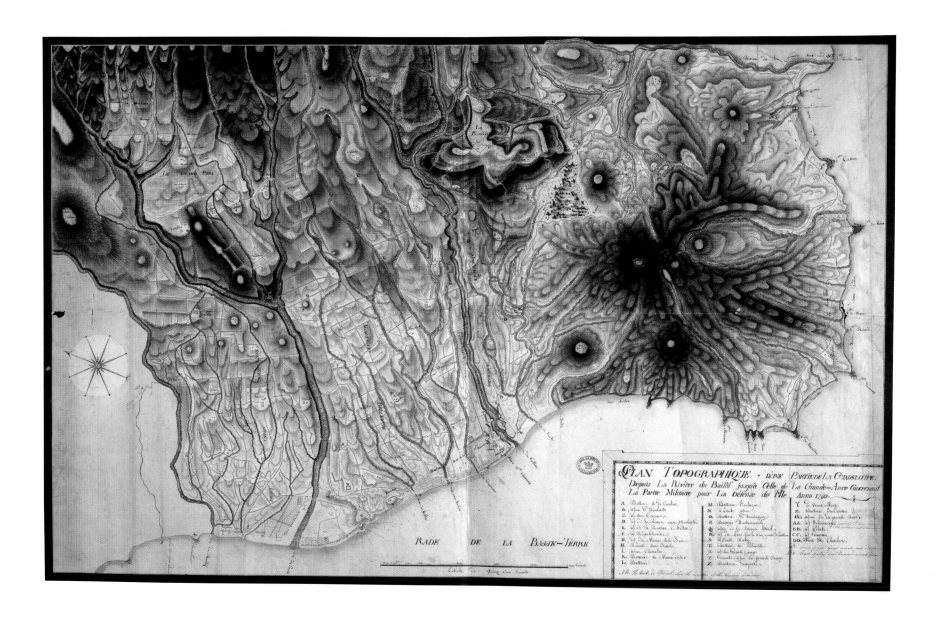

RADE DE LA BASSE-TERRE

# Escape plan  DARTMOOR, DEVON, 1812

This bird's eye view of Dartmoor prison was drawn by someone who clearly knew it well but was apparently unused to his task. He used a strange perspective to impose a detailed site plan onto a watercolour painting of the impressive moorland landscape in which the prison is set, giving a sense of the wild remoteness of its location. A compass indicator at lower right indicates that north is to the left; the viewer faces east, out towards the moors. The prison was at that date newly built, to house French prisoners captured during the Napoleonic wars (c.1803–1815) since facilities on board the prison 'hulks' or derelict ships were known to be inadequate and cramped, with swelling numbers as the war went on. The first prisoners had arrived here in 1809.

The plan gives a good idea of what life in the prison was like. The buildings are lettered and numbered, and explained in a key. The prisoners were kept in the compound on the left, surrounded by iron railings and ramparts, while the guards' barracks are on the right. A note states that the prison was lighted by 146 lamps and the barracks by 20. The prison had some of the services of a small town, with a forge, stables, carpenter, miller and market place. Many prisoners were in poor health, and the prison housed its own doctor and matron, pharmacy and hospital. Outside the prisoners' compound on the left lie the dead house and burial ground.

However, this was not a plan simply made for general reference, but in relation to the matter of an escaped prisoner. In notes at lower right there is mention of 'Mr. V'; papers among which the map was found indicate that this was Louis Francois Vanhille, who had escaped in 1812. The Admiralty was anxious to find out how Vanhille got out, in part to recapture him, but also to punish any local people who had helped him, and in particular to prevent any further escapes. These papers include reports and interviews with local people and a fellow prisoner who knew Vanhille well.

From these we learn that Vanhille was a purser in the French navy, was five feet five inches tall, had a fresh complexion, brown hair, light grey eyes, and was somewhat disfigured by smallpox. He was one of a number of French prisoners who had been living in the nearby Cornish town of Launceston on parole, but he had broken the terms of this arrangement by going to dine with a Dr Mabyn at Camelford without leave, for which misdemeanour he was sent to Dartmoor prison and arrived on 12 December 1811. The next summer 'a young lady of Tavistock' had brought a 'waggoners frock' into the prison for him to escape in disguise. He went first to Cork and thence to Jamaica in the ship *Jane*, calling himself Williams, but was apprehended there by the authorities in February 1813, and sent back to England on the next ship of war.

PRISONER OF F BLOCK: The prison building labelled 15 was where Vanhille was kept, and the blue line shows the limit of the walk he was allowed, as far as the threshold of the main gate to the prisoners' compound.

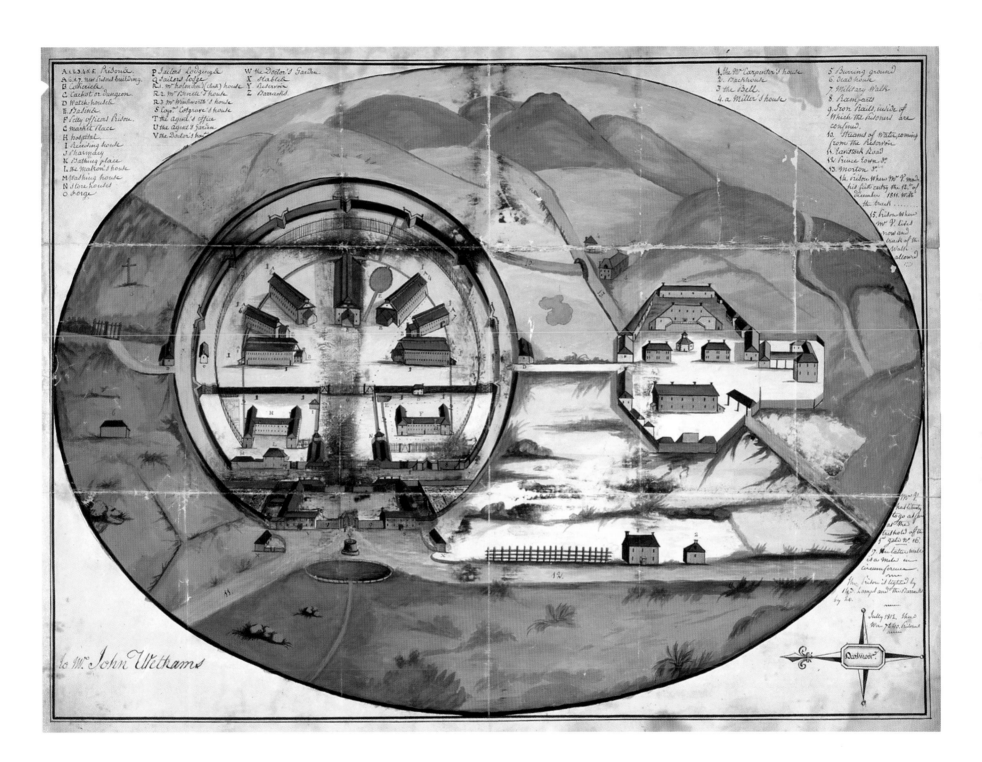

# The obelisk, the pheasantry and the blacksmith

## BOCONNOC, CORNWALL, 1838

When Queen Victoria came to the throne in 1837 a revolution was in progress. For ten centuries clergymen had been supported by the biblical tenth of farm produce, but in 1836 a modernising government passed an Act to change payment in pigs, eggs or corn to more convenient money charges. The result was this map, and nearly 12,000 other tithe maps of parishes in England and Wales. Mostly manuscript, they are often the earliest large-scale maps of these places, made before printed Ordnance Survey maps came into widespread use. Indeed, the Tithe Survey might have become the first national mapping to a consistent scale and standard, and a set of conventional signs was drawn up to this end. However, the cost deterred the government from commissioning such a survey and the burden of providing tithe maps fell upon landowners. This meant that the resulting maps vary widely in style, scale, size, accuracy and amount of detail.

The parish of Boconnoc near Lostwithiel is dominated by a private estate, shown on an unusually attractive watercolour tithe map, part of which we see here. This level of detail was not a requirement of the Tithe Survey, since the estate would be valued in total for the purpose of calculating how much tithe it was due to pay. It seems likely that the map was commissioned by the estate's owner and made by their steward, John Bowen, who also made tithe maps for three neighbouring parishes.

Bowen had given decades of service and would have been very familiar with all the details of the estate. His map shows at left the mansion in red (denoting an inhabited building), with next to it the church in black (uninhabited). There are pleasure gardens, avenues of trees, orchards, a deer park, woods, and the River Lerryn. At some distance is a pheasantry (top left) where pheasants were reared, and to its right an obelisk 123 feet high, erected in 1771. Lands surrounding the estate are shown in less detail. There are named farms, fields, villages, roads and footpaths, and the boundaries of parishes and manors.

As so often with maps in the archives, this map is a key to another record. The plot numbers on the map link to a text document called an apportionment in which details of property owners and occupiers were recorded. In this case, Lady Anne Grenville of Boconnoc House was the main landowner in the parish, and all her tenants are listed, including John Bowen. For his house, garden, meadows and pastures he paid a pound and ten shillings in tithe per annum.

The Tithe Survey was the most detailed study of land use and ownership since Domesday Book over seven centuries earlier, and these records reveal a wealth of information about early Victorian place and society. Many of the maps capture the landscape just before the changes wrought by the railways, industrialisation and urbanisation. Others, such as this one, present a scene little altered from the countryside today.

JOEL PEDLAR'S SMITH'S SHOP: This map unusually gives the name of the blacksmith whose shop lay at a crossroads between two parishes. It was opposite the bowling green – perhaps so that his customers could have a game of bowls while waiting for their horses to be shoed?

Plan of
BOCONNOC PARISH
in the County of
Cornwall
1838

# A-hunting we will go

The British abroad continued many of their home traditions. This map is a product of a fondness for hunting among the upper classes who formed the officer cadre of the army, whether in Scotland, Africa or, as here, in Spain. The British army had garrisoned Gibraltar (off the map to lower right) since 1713, and by 1812 had started the Hunt, when Britain's support for Spain during the Napoleonic Wars made permission to cross the border easy to obtain. British officers and local Spanish gentry together pursued foxes across the countryside north-west of Gibraltar, among the Spanish farms and in the hinterland. High-ranking visitors from England, including royalty, would join or watch the Hunt. What began as a way of keeping military men fit became embedded in the social fabric of the colony.

This is part of a large manuscript 'Hunting Map of the Country Adjacent To Gibraltar'. It shows the area around the Bay of Gibraltar and inland as far north as a remote area of cork-oak woods at Almoraima, a favourite hunting ground (off the top of the map). The use of watercolour heightens the sense of drama in the landscape to the viewer. The map shows how rugged and dangerous this mountainous countryside was for riders galloping at speed, with deep narrow ravines and watercourses to negotiate. The central river valley of the Guadarranque, which runs from the Bay northwards, was both a boon in its flatness and an obstacle, since men and horses alike had to cross the water by a small ferry.

Spanish place names – for towns, villages and haciendas – mix with English ones, for places associated with the Hunt, or to explain types of terrain. These include salt pans in blue on the coast at the bay's centre, sandy hills to their left and a sandy plain around the bay to the right. In the hilly interior dense vegetation and dark green pine woods are indicated. Game bird 'covers' are named for Englishmen with one marked 'Woodcock Cover'. Other sporting details include bullrings, and a steeplechase course above the sandy plain. The lowlands feature plantations, osiers, orange groves, and public gardens by the town of Algeciras (left).

Captain Charles Lacon Harvey made this map while on his second tour of duty in Gibraltar, which finished in April 1873. His army service shows that he was aged 33, and serving in the 71st (Highland) Regiment of Foot. He was trained not in mapmaking but in veterinary science and riding, and also excelled at musketry – all useful skills for a hunter. Harvey went on to serve at the Cape of Good Hope and in the East Indies. His map saw later service, too. The need for this kind of detailed topographical map was shown by the fact that it was printed and used in different versions for decades, both by the Hunt and by the army.

FROM PAINT TO PRINT: This map was transferred to print by lithography to become a monochrome military map published by the Topographical Depot of the War Office in 1874. This lost the immediacy of the original but allowed wider use.

Hunting Map
of the
Country Adjacent to
GIBRALTAR
Surveyed in 1873 By
CAPTⁿ C.L. HARVEY 71ˢᵗ H.L.I.
Scale 2in to 20 or 3in to a Mile.

BAY OF GIBRALTAR

ALGECIRAS

LOS BARRIOS

PALMONES

SALT PANS

EL PRADO

SANDY PLAIN

SPANISH LINES

BRITISH LINES

LINEA

# Seats with a view

If other maps in this chapter focus on a particular place, this railway poster-map is more about a railway company's ability to whisk the late-Victorian traveller from place to place all over the United Kingdom – or at least, to those places where the company had stations and hotels. On a closer look, the map offers enticements to these specific places, tailored to different types of passenger. The businessman or emigrant might be impressed by the huge hotels and ferries at Liverpool and Holyhead. For those wanting a seaside holiday, Morecambe and Llandudno beckon, while Lichfield Cathedral offers culture and the Menai and Britannia Bridges are engineering spectacles. There is much to appeal to connoisseurs of the picturesque at Windermere and Killarney.

The line diagram laid out over the outline map shows far more than the London and North Western Railway Company's own lines. Its coverage extends to other lines over which they had negotiated the right to run their carriages. This allowed them to offer services to the West Country, Wales, East Anglia and Scotland, which were outside the remit of the Company name. It even extends across the Irish Sea by means of the Company's steamships, which delivered passengers to rail services in Ireland. The Company's hotels are listed at the lower edge.

This poster is found among advertisements for ladies' underwear, an opera house, canned tomatoes, pig powders and Christmas crackers, each representing a claim to copyright in the artwork shown. They are each attached to a form giving details of the person or company claiming the rights and, where different, the artist. The form for this map is dated 21 March 1899 and gives the copyright owners as McCorquodale & Co Ltd of London, makers of posters and postcards. This registered poster had the Company's logo trimmed away from the top, in order to fit the form. Complete, it must certainly have attracted attention when displayed on station platforms and in waiting rooms and railway carriages.

The artist is named as Alfred Pernet of 43 Arvon Road, Islington, North London. Little is known of him except that he also painted a Dorset landscape, and that the 1901 census records him as a 'lithographic artist'. It seems likely that he contributed the original watercolour vignettes around the edge of a route map generated by the Company itself. These views are arranged artistically, although places shown in them are nowhere near their actual position on the map. This could be misleading for someone who did not know the country, who might think that Liverpool lay roughly in the position of Dover, for instance.

Land ownership was a major reason for mapping of the countryside, whether for individuals or nations' colonies. The railway companies were major landowners, but of linear rather than consolidated holdings. For them, acquiring land was not an aim in itself, but rather a means to cover that land with an expanding empire of stations, gleaming metal tracks and trains full of passengers enjoying the view.

CREATED BY THE RAILWAY: There was very little at Greenore until 1873, when the London and North Western Railway Company constructed a substantial hotel and railway station to serve ferry passengers. On the southern shores of Carlingford Lough, it offers spectacular views of the Mourne Mountains.

GREENORE, County Louth, Ireland.

# CHAPTER FOUR

# THEATRES OF WAR:
# military maps

Conflict is a recurrent theme throughout history, and over the centuries it has been an important influence on the development of cartography. Maps relating to defence and warfare are to be found throughout this book. We have also mentioned the long tradition of British military mapmaking and its fine quality in previous chapters, notably on pages 63 and 93. In this chapter, we explore how maps can shed light on battles, campaigns and other aspects of war. Unsurprisingly, most of our selected maps are of military origin. They tend to be drawn clearly and precisely, reflecting the discipline of service life; this is frequently in striking contrast to the violent, sometimes chaotic, events that they portray.

Our focus here is upon land-based warfare and the army, and later also upon conflict in the air. Chapter 5 is devoted to maps (or rather charts) of the sea. Nevertheless, many operations have relied upon the combined efforts of soldiers and sailors, and so aspects of naval warfare have crept into this chapter. Most of the maps that we have chosen

represent wars in which British troops were actively involved, and the majority were made by British service personnel. The curious tale of why our plan of the 1663 siege of Neuhäusel is an exception to both of these is outlined on page 111.

The changing pattern of political alliances within Europe is a theme that unites much of this chapter. The French, so often enemies of the English throughout history, feature heavily. The old rivalry between Great Britain and France – both in Europe and later in colonial North America – reached its zenith during the Napoleonic Wars of the early 19th century, but then subsided. In the 20th century it was replaced by the alliance known as the Entente Cordiale. By contrast, Germany – a region which had traditionally supplied British royalty with suitable Protestant marriage partners – would become the United Kingdom's chief enemy during the two world wars. Military action outside Europe included armed conflict with people who resisted British colonisation of their lands;

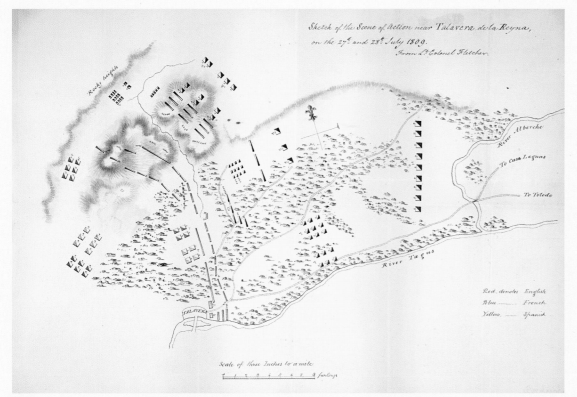

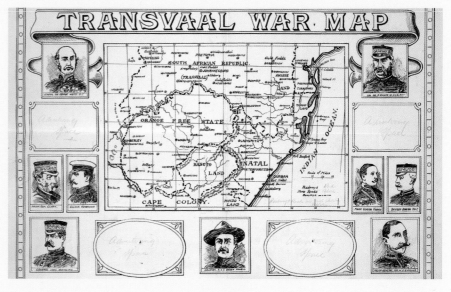

**1** BATTLE PLAN: The Peninsular War of 1808–1814 was one of the series of conflicts known collectively as the Napoleonic Wars. This map records the nominal victory of the British (in red) and Spanish (yellow) over the French (blue) at the Battle of Talavera on 27–28 July 1809.

**2** NEW FORTRESS: Built by the English at the beginning of the 17th century, Charlemont Fort in County Armagh, Ireland, was an active military site for more than 150 years, and was famously besieged in 1650. Little of it now survives.

**3** COMMANDING THE TROOPS: This map was probably designed as a show of support for the British during the Second Boer War of 1899–1902. Among the portraits of commanders forming its border is Robert Baden-Powell (bottom centre), who later founded the Scout Movement.

on page 123 we look at a battle from the New Zealand Wars.

Maps have played many roles in military life. They are vital tools when preparing for war, whether in the course of building or extending defence works (as shown on page 115), or during reconnaissance to establish the lie of the land. On page 119 we discuss a proposed attack that was never carried out. During military engagements, maps may of course be used for directing the action, but this is not their only function. Combatants and other witnesses have made maps to record or report upon the progress or results of operations and battles; pages 109, 117 and 127 supply three very different examples. In practice, maps of this kind are generally drawn shortly after, rather than in the midst of, the events to which they refer.

More decidedly within the category of maps made after the fact are those intended to aid later assessment or reassessment of the reasons for a victory or defeat. On page 121 we feature an example produced to help a court determine the merits of a libel case. More commonly, however, the evaluation was a matter for the military wing of government; this is the origin of the maps on pages 125 and 131. The decorative map of the United Kingdom opposite offers a different kind of retrospective view: the commemoration of past conflicts and the celebration of military successes.

It is common for maps that were originally created at one stage of the cycle of war and peace to have been used or re-used at a different stage, perhaps during a subsequent conflict. The British War Office retained many plans of fortifications and past battles to support the planning and conduct of future campaigns. In a broader sense, this opportunity to re-examine historic military successes and failures through maps and other documentation continues today through their preservation in the archives. One of the primary reasons for the existence of archives is to allow people in the present and future to learn from the experiences of those who lived and died in the past.

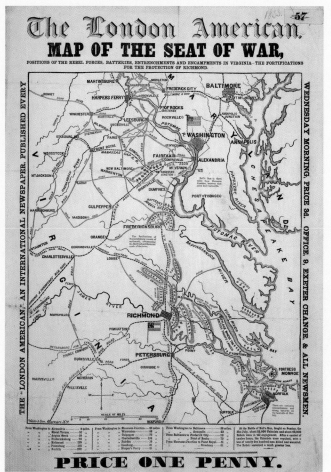

**4** BATTLE HONOURS: Produced during the Second World War, this poster expresses pride in the historical military achievements of the British army.

**5** READ ALL ABOUT IT!: Printed and sold in London in 1861, this map helped to keep the British public informed of events in the American Civil War. It also served to advertise a newspaper that supported the northern, anti-slavery states.

# Saints and sinners SAINT-DENIS, FRANCE, 1567

France during the late 16th century was a nation rocked by religious conflict. The majority of its population were Roman Catholics, but a substantial minority – around two million people, drawn from all levels of society – had embraced Calvinist Protestantism. The Huguenots (as the Calvinists became known) felt that intolerance from the Catholic establishment denied them the freedom to practise their faith. Many Catholics, particularly among the upper classes, saw Protestantism as a threat to the nation's cohesion and stability. A sequence of civil wars and attacks, punctuated by uneasy truces, lasted from 1562 to 1598. Under the Edict of Nantes, King Henri IV, himself a former Protestant who had converted to Catholicism, brought about a more lasting peace by granting the Huguenot community a degree of officially-sanctioned religious liberty.

This manuscript bird's-eye view portrays one of the bloodiest engagements of these wars, the Battle of Saint-Denis, near Paris. On 10 November 1567 a 16,000-strong army of Catholics commanded by Anne de Montmorency fought a much smaller Protestant force under Louis de Bourbon, Prince de Condé. Despite being outnumbered by more than four to one, the Huguenots held out for several hours before conceding defeat and retreating to the east. The 74-year-old Montmorency was severely wounded and died two days after his victory.

As much a picture as a battle plan, the sketch is not easy to interpret. It offers no firm impression of the progress of the fighting and does not even indicate clearly which troops belong to which side. It is possible that it employs an unusual time-lapse effect, with the right-hand side showing events earlier in the battle than the left-hand side. The geographical position of the battlefield, however, is firmly anchored by the views of cities and towns depicted at its fringes: Saint-Denis to the north, Aubervilliers to the east, Paris and La Chapelle to the south, and Saint-Ouen to the west. Disciplined ranks of infantrymen march across the lower half of the sheet, while in the upper half a charge of cavalry sweeps from right to left. To the right of the centre cannon are shown being loaded and fired.

The plan is believed to be the work of three young Englishmen – Edward Berkeley and two of the sons of Sir Henry Norris, the English ambassador – who were then resident in Paris and may have witnessed some of the battle. Norris enclosed it with a despatch sent a few weeks later on 29 November, one of his frequent reports to Queen Elizabeth I and her senior ministers about events in France. A skilled political operator, Elizabeth was keen to gather intelligence about conflicts and intrigue in mainland Europe. Having herself experienced life as a Protestant in a Catholic country during the reign of her sister and immediate predecessor Mary I, she was also acutely aware of the potentially deadly implications of not appearing to conform in matters of religious faith and practice.

MORTALLY WOUNDED:
This skirmish may depict Montmorency being knocked from his horse.

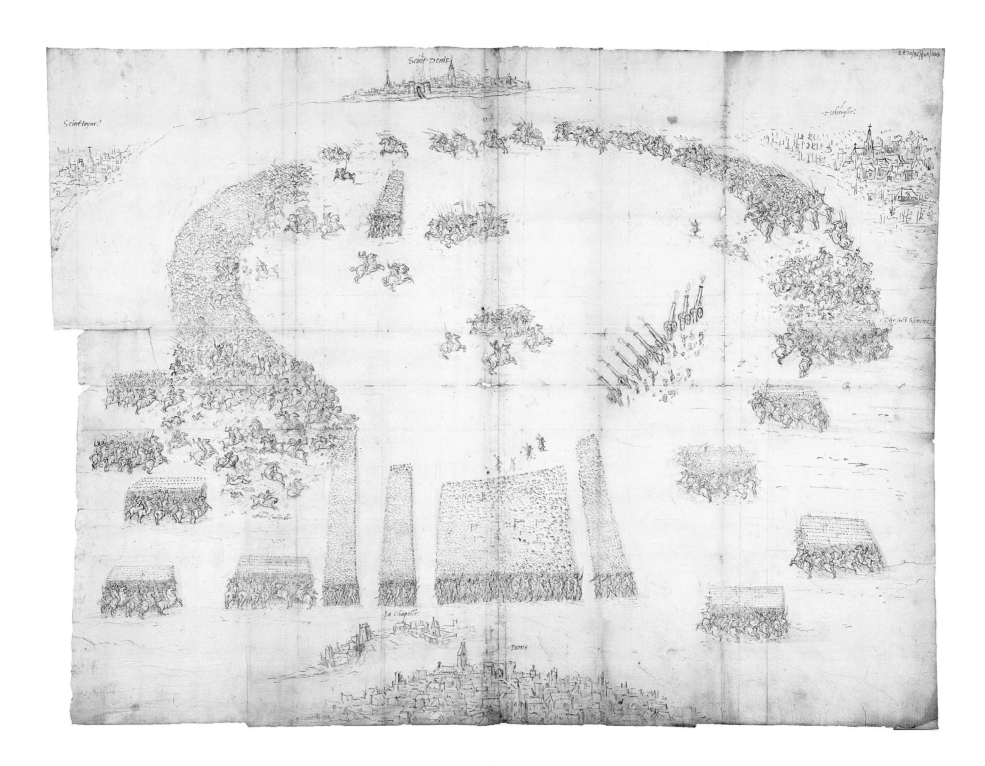

Seint Denis

Seint toyne

Esterngtte

Chemy Kontes

La Chapelle

Paris

# A siege and a spy

NEUHÄUSEL (NOVÉ ZÁMKY), KINGDOM OF HUNGARY, 1663

Many towns and cities within central and eastern Europe have been known by several different names over the centuries, as successive wars and treaties have shifted international boundaries back and forth. Today the town of Nové Zámky lies in Slovakia, but for a long time it was part of the Kingdom of Hungary and called Érsekújvár. On this map, it is labelled with yet another name, the German Neuhäusel. All three of these names embody the meaning of 'new castle', a reference to the town's traditional status as a defensive point in a border region. The star-shaped fortress near the centre of the map was built in the 1570s, its pointed walls designed to withstand the type of sustained cannon fire that is depicted here.

For much of the 16th and 17th centuries, what had formerly been unified Hungary was divided between the Austrian Hapsburg family, who ruled over the west, and the Ottoman (or Turkish) Empire, which had conquered the east. The boundary between these two rival states fluctuated. For most of this era Érsekújvár was within the Hapsburgs' domain but it spent several periods under Ottoman control. The month-long siege of the town that is portrayed here took place in August and September 1663, and ended with a victory for the Turkish army. The region remained in Ottoman hands until 1685.

The map is a woodcut – an image printed from a carved block of wood – attributed to Georg Lackner. To modern eyes, its bold, stylised lines resemble the work of a cartoonist. It is not drawn to scale: the oversized and imposing tents of the besieging army seem to overwhelm the fortress. The text around the edges of the map includes the names of senior Ottoman commanders. Meanwhile, on the right-hand side, reinforcements arrive to join their comrades.

Our print of the map has an unusual history. It was once owned by Colonel John Scott, an English-born rogue whose colourful career included periods as a sailor, a fraudster, a mapmaker, a fur trader in North America and a soldier in the Dutch army. Frequently in trouble with the law, he was imprisoned several times but regularly escaped punishment for his misdemeanours. He is best known, however, for his work as a spy. He may have been a double or triple agent, employed by the French and Dutch governments as well as by his own country.

Scott's map came into the government's hands on 18 May 1682, when a messenger named Thomas Atterbury confiscated it, along with other papers from his rooms in London. Scott was accused of using his cartographic skills against English interests by making plans of ports with notes about their defensive capability. Letters found in his possession were considered evidence that he had conducted 'dangerous correspondence with foreign powers'. As so often in his past, Scott survived this incident relatively unscathed. He eventually settled in the West Indies, where he died in 1704, aged about 72.

FLYING THE FLAG:
The banners of the besieging forces bear a crescent moon, a symbol of both the Ottoman Empire and its Islamic faith.

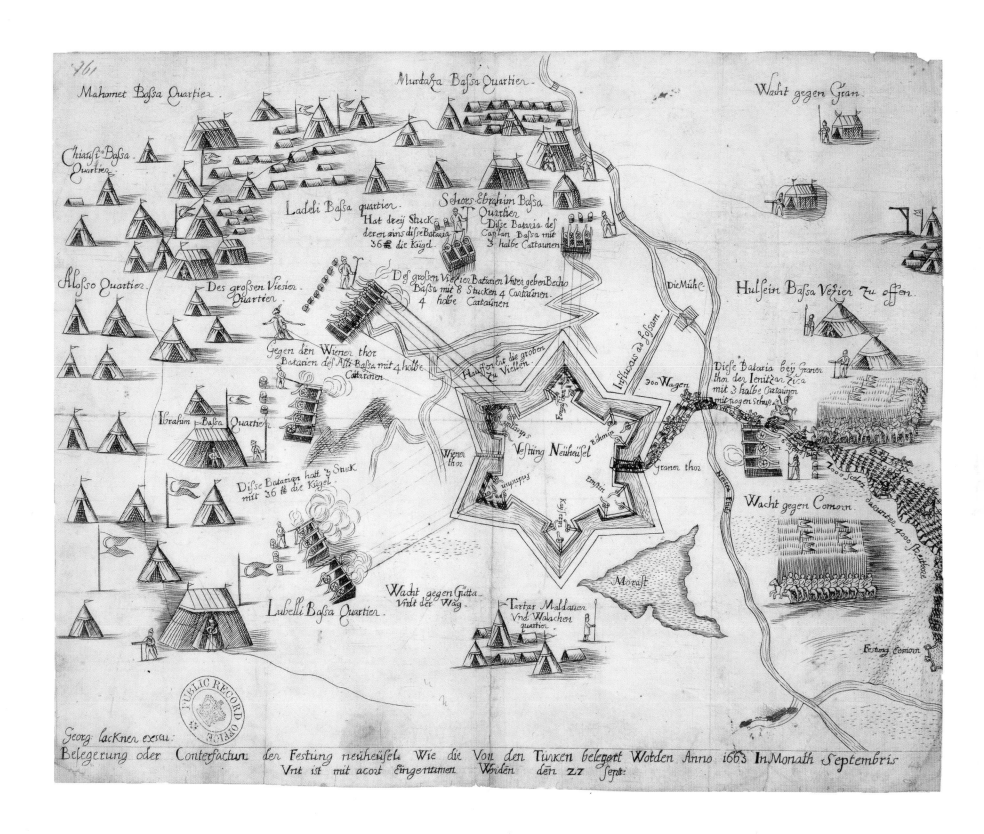

761

Mahomet Bassa Quartier.

Murtaha Bassa Quartier.

Wacht gegen Gran.

Chiausi Bassa Quartier.

Ladeli Bassa quartier.
Hat drey Stuck deren ains disse Bataria 36 ℔ die Kügel

Sciors Ebrahim Bassa Quartier
Disse Bataria des Caplan Bassa mit 3 halbe Cattaunen

Alosso Quartier.

Des großen Viesier Quartier.

Def großen Viesier Baturien Vnter geben Bechio Bassa mit 8 Stucken 4 Cartaunen 4 halbe Cartaunen

Hulsein Bassa Vesier zu offen.

Gegen den Wiener thor Batarien des Asti Bassa mit 4 halbe Cattaunen

Die Mühl

Influxus ad fossam

200 Wagen

Diese Bataria beÿ Graner thor der Ienitzar Zica mit 3 halbe Cartaunen mit wagen schuß

Ibrahim Bassa Quartier.

Haußer vet die groben zu Viellon.

Vesting Neüheüsel

Disse Batarien halt 3 Stuck mit 36 ℔ die Kügel

Wiener thor

Wacht gegen Comorn.

Wacht gegen Gutta Vnd der Wag.

Tartar Maldauer Vnd Walachen quartier

Morast

Festung Comorn

Lubelli Bassa Quartier.

Georg lackner excu:

Belegerung oder Conterfactur der Festung neüheüsel Wie die Von den Turken belegert Worden Anno 1663 In Monath Septembris
Vnt ist mit acort Eingenumen Worden den 27 Sept:

# George Washington – surveyor

As the first President of the United States of America, George Washington is one of the most famous figures in history. This map illustrates one of the lesser-known episodes of his life. By the mid-18th century, the longstanding rivalry between Great Britain and France had extended from Europe to North America, as each country sought to colonise swathes of the new world. In the autumn of 1753, Robert Dinwiddie, the British lieutenant governor of Virginia, was alarmed by reports of French advances southward from Lake Erie to the Ohio Valley. Anxious to assert British claims to the territory covered by this map – which corresponds roughly to the western fringes of modern-day Pennsylvania – he wrote a formal letter demanding the withdrawal of French troops from the region. He appointed the 22-year-old Washington, then a major in the Virginia militia, to deliver it.

Washington's party crossed the Allegheny Mountains in mid November and proceeded north-west to the Forks of the Ohio. This point (near the centre of this map) is the confluence of the Monongahela and Allegheny Rivers; the latter is labelled here as the Ohio River. He then spent several days at nearby Logstown, in discussions with two Native American leaders, the Seneca Tanaghrisson (known to colonists as 'the Half-King') and the Oneida Monacatoocha. Both men shared British concerns about French expansion. Delayed by bad weather, Washington eventually reached Fort Le Bœuf (shown at upper right at the head of French Creek) on 11 December. His meeting with the French commander, Jacques Legardeur, was cordial but not a success. Neither side was willing to give up any claim to the region. Legardeur also disavowed any authority to negotiate on behalf of his superiors, who were based in faraway Quebec City.

On his return to Virginia the following January, Washington submitted a detailed report about his activities to Dinwiddie, who decided to send a copy of it to London (along with this map) and to have it published. These actions by the governor brought Washington to wider public notice for the first time. This initial taste of fame, along with the experience gained during his spell in the British militia, would stand him in good stead when he came to lead the fight for American independence.

Before receiving his commission in the militia, Washington had worked as a land surveyor. He was therefore a competent draughtsman and is believed to have drawn this map himself. A note on it records his recommendation to build a British fort just east of the Forks of the Ohio, to guard against further French settlement in the area. Work on the new Fort Prince George began early in 1754, but it was overrun a few months later by the French, who replaced the unfinished British construction with the larger Fort Duquesne. This in turn was destroyed by the British in 1758, and replaced with Fort Pitt. The site is currently occupied by Point State Park in downtown Pittsburgh.

FORKS OF THE OHIO: George Washington suggested this strategically important site for a British fort because it commanded the confluence of the Allegheny and Monongahela Rivers.

STARS AND STRIPES: The medieval seal of the Wassyngton family – English ancestors of George Washington – is thought to have inspired the design of the United States flag.

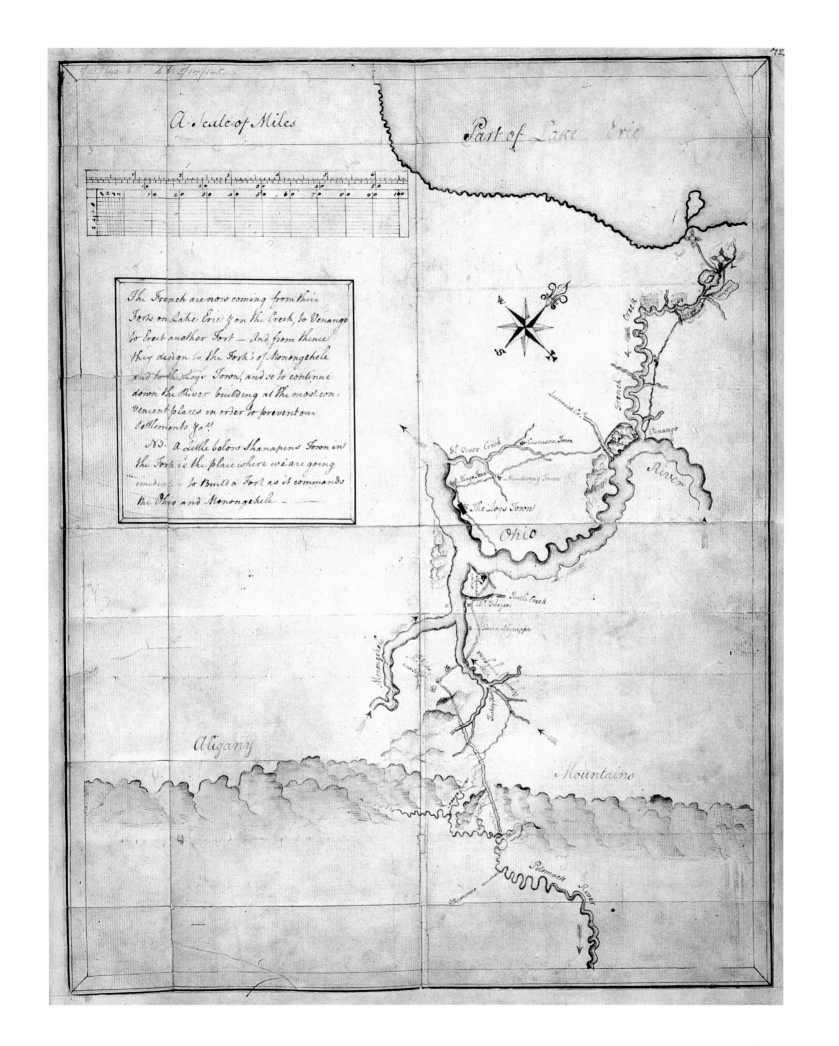

A Scale of Miles

Part of Lake Erie

The French are now coming from their
Forts on Lake Erie & on the Creek, to Venango
to Erect another Fort — And from thence
they design to the Forks of Monongehele
and to the Logs Town, and so to continue
down the River building at the most convenient places in order to prevent our
Settlements &c.

NB. A little below Shanapins Town in
the Fork is the place where we are going
immediately to Build a Fort as it commands
the Ohio and Monongehele —

Gr. Bever Creek    Kuscusca Town    Laemiahtin.    French    Creek    Venango

Kinga Town    Mandering Town    River

The Logs Town

Ohio

Turtle Creek

Queen Aliquippa

Monongehele

Aligany    Mountains

Potomack    River

# The greatest fortification in Europe

For more than five centuries, the history of the Balearic Islands in the western Mediterranean Sea has been closely linked with that of mainland Spain, with one notable exception. Minorca, the second largest island in the archipelago, passed much of the 18th century as a British colony. The island was captured by a joint Anglo-Dutch force in 1708, during the War of the Spanish Succession. Work to improve its fortifications began almost immediately, even before British sovereignty was regularised under the Treaty of Utrecht in 1713. These defence works were concentrated on the eastern side of the island, around the port of Mahon, which under British rule replaced the old city of Ciudadela as Minorca's capital.

Chief among the fortifications was St Philip's Castle, designed to protect Mahon's fine harbour, which lies off this map to the right. The diamond-shaped core of the fortress had been built by the Spanish in the mid-16th century and enlarged somewhat during the 17th century. Nevertheless, the castle as seen here – an extravagant flowerlike shape formed of counterguards and ravelins, surrounded on the landward side by a semi-circle of batteries, lunettes and redoubts – was largely the result of work conducted by the British at great expense during the 18th century. The maze of underground galleries and passages (shown coloured grey) were of mixed Spanish and British origin.

By 1754, when this plan was drawn by Thomas Souter, the castle was the largest of its kind in Europe and was still growing. The stout walls and rugged cliffs depicted here present a conscious impression of strength and impregnability, reinforced by the arms of King George II (at left) and a suitably military cartouche (at upper right) featuring weaponry, drums and a snarling beast. Even the sea looks threateningly unsettled.

When put to the test, Minorca's defences proved rather more vulnerable than the plan suggests. In 1756, at the beginning of the Seven Years War, the French successfully besieged the castle and gained control of the island. In a controversial incident, Admiral John Byng, who had commanded the British fleet off Minorca, was court-martialled, convicted of failing in his duty and subsequently executed. This episode was memorably satirised by the French writer Voltaire in his novella Candide, in which the Royal Navy is said to have shot one admiral 'to encourage the others'.

As conflict in Europe continued intermittently over the following decades, Minorca spent two more periods in British hands, from 1763 to 1782 and from 1798 to 1802. St Philip's Castle, however, never regained its former glory. It was destroyed in 1782 on the orders of the Spanish King Charles III after his country regained control of the island. Another fort was later built on the site, but this too was demolished in 1805. Although almost nothing of the once-proud fortress remains above ground, several of the subterranean galleries survive, and the outline of the castle is still clearly visible from the air.

BUILT FOR A PURPOSE: The plan records the uses of many parts of the castle, from officers' quarters to a carpenter's shop, and from powder magazines to stores for beef and wine. This building was designed as a hospital.

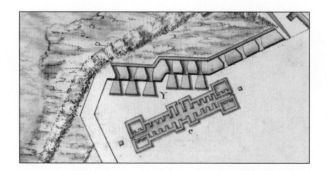

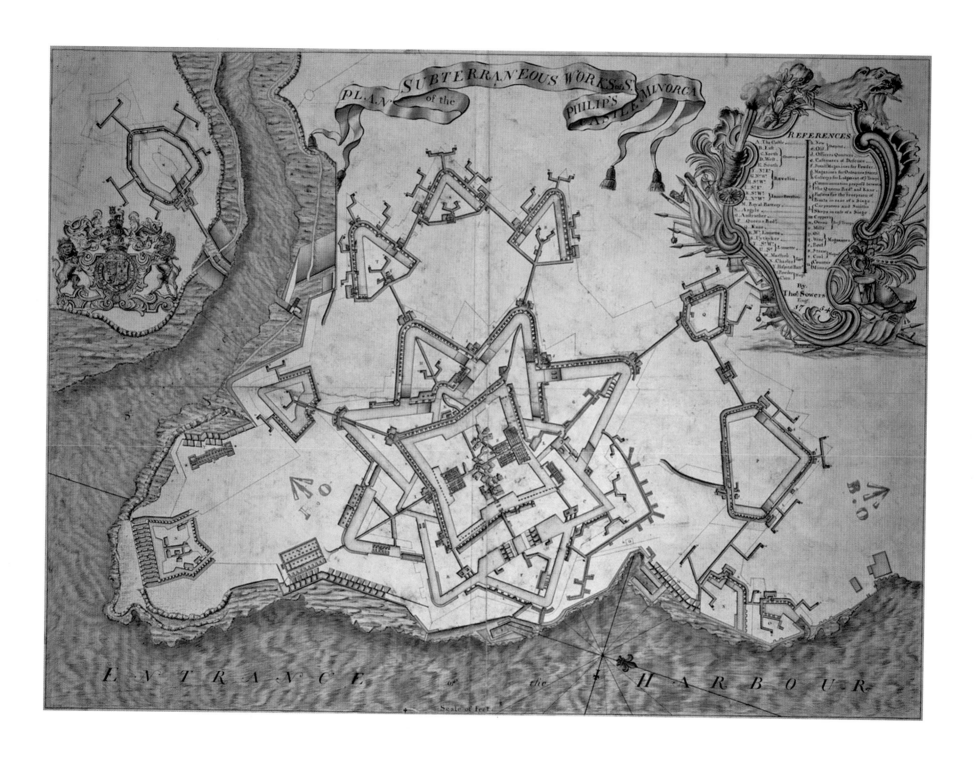

PLAN of the SUBTERRANEOUS WORKS of St PHILIP'S CASTLE MINORCA

REFERENCES

By Thos Sowers

ENTRANCE of the HARBOUR

Scale of feet

# 'The best fort in America' NIAGARA, 1759

The Seven Years War was one of the first truly global conflicts. Although driven chiefly by European power politics, battles took place in Asia, Africa and the Americas, as well as within Europe itself. In fact, it was in North America (where swathes of territory were disputed between French and British colonists) that its first blows were struck in 1754. Fighting would not begin in Europe until 1756, when a new set of alliances – Austria, France and Russia ranged against Great Britain and Prussia – was established. The North American theatre is often known by the separate name of the French and Indian War, a designation that recognises the role played in the conflict by Native Americans, many of whom fought alongside either the British or the French.

1759 was a year of military success for Great Britain, and the Battle of La Belle-Famille on 24 July (shown here at lower right) was a particularly well-executed operation. A British contingent under Lieutenant-Colonel Eyre Massey (positions coloured gold) ambushed French troops (coloured mauve), to prevent them from joining their compatriots at Fort Niagara (point A). In a letter to William Pitt the Elder, who was the British government minister responsible for the American colonies, Massey claimed that his men had chased the French for seven miles through the woods. Many French officers were captured, including their commander, François-Marie Le Marchand de Ligney, who subsequently died of his wounds.

Massey's motive for writing directly to Pitt was to ensure that he would be credited for his contribution to the capture of Fort Niagara, which surrendered to the British two days later. He was careful to explain that the French had lost 'the best fort in America' as a direct consequence of his actions, and he enclosed this map in his letter to illustrate the point. Drawn with military precision, although the scale is unspecified, it clearly lays out the geography of the area. The flow of the Niagara River northward to Lake Ontario is marked, as is the road leading south to the famous waterfalls. Point C, at the northern end of this road, was a burial ground, presumably used to inter the casualties of this skirmish. The garden plot depicted on the western (now Canadian) side of the river provides an uncharacteristically bucolic touch, as it seems unlikely to have held any strategic significance.

The eventual outcome of the Seven Years War mirrored that of this battle: defeat for the French and their allies. Under the Treaty of Paris in 1763, France lost all of her former territories on the North American mainland to Great Britain, with the exception of western Louisiana, which she had already secretly ceded to Spain. The British maintained control of Fort Niagara throughout the American Revolutionary War of 1775–1783, and for some time afterwards. Although the boundaries agreed in 1783 gave the fort to the newly-independent United States, only in 1796 did it actually pass into American hands.

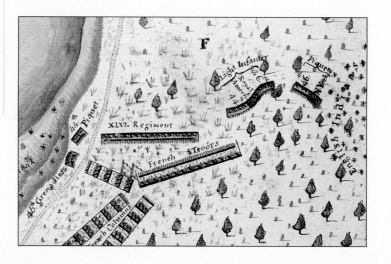

'ENGLISH INDIANS': The map acknowledges the contribution made by Great Britain's Iroquois allies to her victory in this battle.

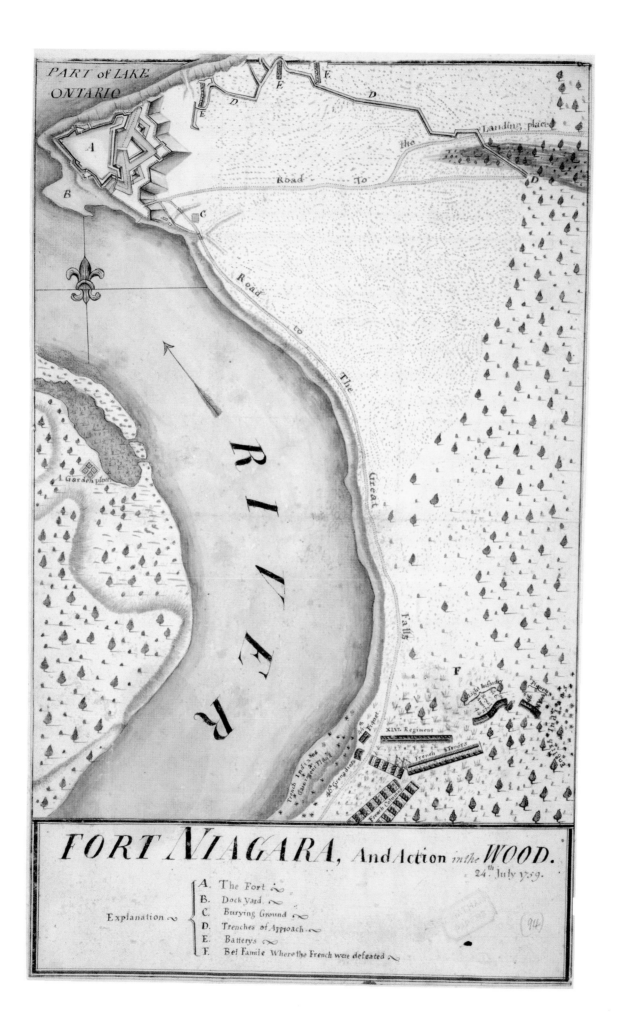

PART of LAKE ONTARIO

Landing place

Road To the

D

Road to The Great Falls

R I V E R

A Garden plott

Light Infantry

Picquets

XLVI. Regiment

French Troops

French 1st Ser

Grenadiers

FORT NIAGARA, And Action in the WOOD.
24th July 1759.

Explanation

A. The Fort
B. Dock Yard
C. Burying Ground
D. Trenches of Approach
E. Batterys
F. Bel Famie Where the French were defeated

(94)

# Balloon debate   BREST, FRANCE, c.1800

Since ancient times, warfare has been conducted both on land and at sea. Regular combat in a third environment – the sky – is of much more recent origin. This map, dating from around the turn of the 19th century, incorporates aspects of all three modes of conflict. It depicts both the harbour of Brest in western Brittany, a major Atlantic-facing base for the French navy, and a number of defences on land, notably the fortifications of the city of Brest itself. The map's most striking feature, however, is the six balloons shown hovering over the water. Three of these are dropping explosive devices upon French ships within the harbour. The others sit poised above the open sea, taking their bearings from the British vessels at lower left in preparation for a further attack.

The raid portrayed here (and described briefly in an accompanying note) was a purely hypothetical operation. Great Britain did not actually launch any airborne attacks on France or other countries during this period. This particular scheme was the brainchild of Charles Rogier, a dancing master who lived in Chelsea, then on the outskirts of London. The son of an English mother and French father, he considered himself a patriotic Briton and was a staunch supporter of his country's armed forces. He was also a keen amateur inventor whose other proposals included: a lighthouse, a whaling harpoon, a scheme for defending Gibraltar with boiling water, a noiseless carriage, and a scheme for relieving the national debt.

Although certain details are Rogier's plan are fanciful – particularly his suggestion that the explosives could be discharged by clockwork – most of the technology required for it lay well within the bounds of possibility. Owing to the efforts of the Montgolfier brothers and other pioneers, mainly in France, human pilots had successfully flown balloons powered by both hot air and hydrogen during the 1780s. The French army had been quick to see the possibilities of ballooning. The earliest recorded use of aeronauts in European warfare was at the Battle of Fleurus in 1794, although there the victorious French employed their balloon for reconnaissance purposes and not for launching the 'spiked Rockets' or 'Chymical, inflammable Liquids' that Rogier later envisaged.

The British did in fact deploy rockets against the French during the Napoleonic Wars, but fired them only on land or from boats, never from the air. Nonetheless, the Secretary of State for War and his staff apparently took Rogier's idea seriously enough to retain this map for reference. The French Revolution and the subsequent regimes of the Republic and the Emperor Napoleon had prompted bitter conflict between France and Great Britain. In this political context, the idea of using ballooning technology against the nation which had perfected it must have been highly appealing to British leaders and officials. It is their foresight that we have to thank for the survival of this map, the oldest in the archives to show a military operation in the skies.

ROCKET LAUNCHER: A balloon discharges its weapons, causing a spectacular conflagration on one of the hapless vessels below.

# 'Forward, the Light Brigade!' BALAKLAVA, RUSSIAN EMPIRE, 1854

The Crimean War of 1853–56 saw Russia pitted against a coalition of the British, French and Ottoman Empires and Piedmont-Sardinia. Often regarded as the first 'modern' war, it featured improved support services, such as the transformation of military nursing brought about by Florence Nightingale and others, and the use of railways to move troops. Events on the front line were also subject to greater scrutiny than ever before. The electric telegraph enabled senior officers to communicate with their governments and war correspondents to send reports to their newspapers. Alongside the more traditional media of sketches, maps and the written word, photography was used to capture people, places and events.

Although Russia's opponents attacked her on several fronts, including her Baltic and Pacific coasts, the most important engagements were concentrated around the Black Sea, with a particular focus on the Crimean Peninsula, which lent its name to the war. The coalition spent nearly eleven months trying to capture the strategic Russian naval base of Sevastopol, which they eventually achieved in September 1855. This foreshadowed Russia's eventual defeat the following February.

The Battle of Balaklava on 25 October 1854 achieved notoriety because of the skirmish portrayed on this hand-drawn map, the Charge of the Light Brigade. Major-General James Brudenell, the Earl of Cardigan, led a 600-strong force of lightly-armoured cavalry into the valley between the Fedyukhin Heights and Causeway Heights. Having misunderstood an order from Lord Raglan, who was in command of the British at Balaklava, Cardigan expected to pursue some retreating enemy forces. Instead, the Light Brigade faced heavy fire from the Russian artillery. The map shows how British forces (marked in pink) swept diagonally from lower left to upper right along the dashed line until they encountered the Russians (marked in green). Large numbers of British men and horses were killed, and many others were seriously injured or taken prisoner.

The poet Alfred (later Lord) Tennyson famously represented the Charge as a heroic defeat in the 'Valley of Death' – an evocation mirrored by the rugged and menacing landscape depicted here. Contemporary public opinion largely followed Tennyson and saw Cardigan as 'the hero of Balaklava'. By contrast, his immediate superior the Earl of Lucan (also his brother-in-law and rival), shouldered much of the blame for the disaster.

Not everyone shared this view. Colonel Somerset Gough Calthorpe (who had been one of Lord Raglan's aides-de-camp) published a book that presented a negative assessment of Cardigan's actions in Crimea. In 1863 the outraged Cardigan unsuccessfully sued Calthorpe for libel in the Court of Queen's Bench. This map was used to illustrate two of the many affidavits that were presented to the court. Colonel George Wynell Mayhow and Captain Daniel Hugh Clutterbuck – both survivors of Balaklava – supported Calthorpe's version of events. Although more than eight years had passed since the Charge, it was still sufficiently fresh in their memories that they could swear that the map represented it accurately.

'INTO THE VALLEY OF DEATH': This detail from a view of the landscape shows the Charge of the Light Brigade in the background.

"A"

MAP TO ILLUSTRATE

# THE CHARGE OF THE LIGHT CAVALRY AT THE BATTLE OF BALAKLAVA OCT. 25, 1854.

This is the Map ... referred to in the affidavit of George Wynell Mayow sworn before me this 30th day of May 1883

This is the Map ... referred to in the affidavit of Daniel Hugh ... sworn before me this ... day of May 1883

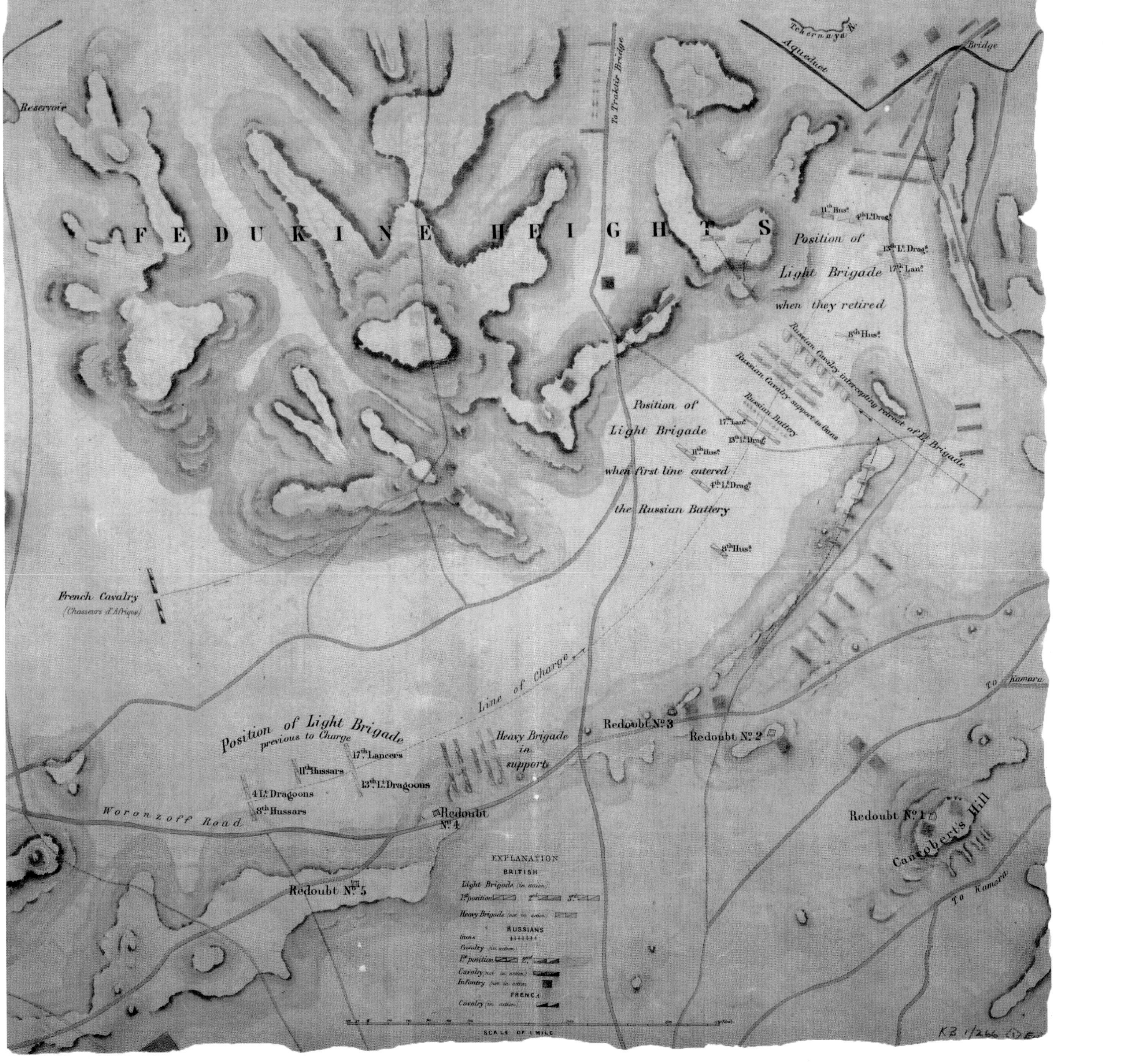

Reservoir
To Traktir Bridge
Tchernaya R.
Aqueduct
Bridge
FEDUKINE HEIGHTS
Position of Light Brigade when they retired
11th Hus.
4th Lt. Drag.
13th Lt. Drag.
17th Lan.
8th Hus.
Russian Cavalry intercepting retreat of Lt. Brigade
Russian Cavalry support to Guns
Russian Battery
Position of Light Brigade when first line entered the Russian Battery
17th Lan.
13th Lt. Drag.
11th Hus.
4th Lt. Drag.
8th Hus.
French Cavalry
(Chasseurs d'Afrique)
Line of Charge
Position of Light Brigade previous to Charge
17th Lancers
11th Hussars
13th Lt. Dragoons
4 Lt. Dragoons
8th Hussars
Heavy Brigade in support
Redoubt No. 3
Redoubt No. 2
To Kamara
Woronzoff Road
Redoubt No. 4
Redoubt No. 1
Canrobert's Hill
Redoubt No. 5
To Kamara
EXPLANATION
BRITISH
Light Brigade (in action)
1st position 2nd 3rd
Heavy Brigade (not in action)
RUSSIANS
Guns
Cavalry (in action)
1st position 2nd
Cavalry (not in action)
Infantry (not in action)
FRENCH
Cavalry (in action)
SCALE OF 1 MILE

# Ambush at the ford

After the United Kingdom had signed the Treaty of Waitangi with chiefs of the indigenous Maori people in 1840, the British regarded the whole of New Zealand as a colony within their empire. Many Maori saw the situation differently. Disagreements over the treaty's implications for their rights to land sparked a sequence of conflicts now known collectively as the New Zealand Wars, which took place intermittently between 1845 and 1872. Not all Maori actively resisted colonisation; some even joined the 'British' side to fight alongside local militiamen and members of the regular army. Those Maori who opposed colonial expansion – considered 'rebels' by the British – were skilled combatants and creative strategists, but were eventually overcome by the weight of superior numbers and firepower.

This map depicts an incident that took place on 11 February 1864, during the Waikato War. Maori forces launched a surprise attack on some British troops who were bathing at a ford in the Mangapiko Stream (shown at lower right). The Maori position is marked with dark blue dots in the scrub inside the curve of the stream. Reinforcements were called in – the great Maori fortress of Paterangi (at top left) and a British camp (at left) were both nearby – and soon several hundred men were fighting on each side. Six British soldiers and about 28 Maori were killed.

Although this was not a major battle, it was brought to wider notice because of the actions of Charles Heaphy, a major in the Auckland Militia. He rescued an injured soldier under intense fire – so heavy that 'Five balls pierced his clothes and cap' – and continued to help wounded men, despite being badly hurt himself. As a result of his actions, Heaphy was awarded the Victoria Cross, the highest gallantry medal for members of the British armed services. He was both the first colonial soldier and the first non-regular soldier to earn this honour. This and another map, both drawn by Heaphy himself, were included in a dossier of evidence submitted to the War Office in support of his claim to the medal.

Heaphy's father, Thomas, was a talented painter who had served the Duke of Wellington as an artist during the Peninsular War. Charles also trained as an artist himself, at the Royal Academy in London. In 1839, aged about nineteen, he became a draughtsman working for the New Zealand Company, which set out to colonise those islands. For much of his career, he worked for the colonial government in various roles connected to land administration, including the surveying of lands taken from Maori after the wars. He also served for a time as a member of the New Zealand House of Representatives and as a judge in the Native Lands Court. Heaphy's official career, however, was undistinguished compared to his artistic achievements, and he is best remembered for his fine topographical views. This beautifully drawn map reflects his skill as a draughtsman no less than his bravery as a soldier.

'FOR VALOUR': The Victoria Cross was instigated during the Crimean War. It is awarded for acts of conspicuous bravery in the face of the enemy.

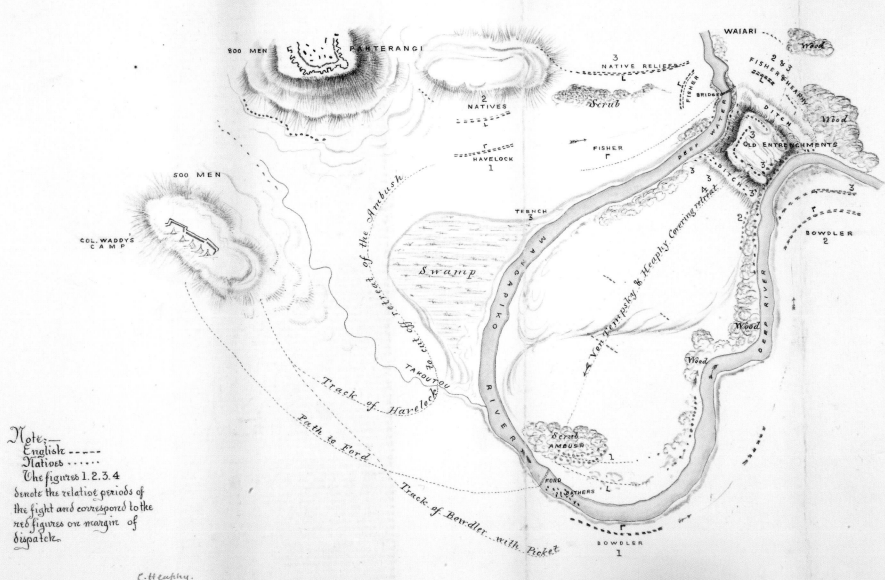

141

800 MEN    PAHTERANGI

NATIVES 2

NATIVE RELIEFS 3

WAIARI

Wood

FISHER & HEARN 2 & 3

Scrub

FISHER

BRIDGE

Wood

DITCH

OLD ENTRENCHMENTS 3

HAVELOCK 1

FISHER

500 MEN

COL. WADDY'S CAMP

DEEP WATER

DITCH 3

3

4

3

2

BOWDLER 2

TRENCH 3

Line of the retreat of the Ambush

A. Von Tempsky & Heaphy Covering retreat

Swamp

WAIKATO RIVER

DEEP RIVER

Wood

Wood

TAKOUTOU

Track of Havelock

Scrub AMBUSH 1

Path to Ford

Track of Bowdler with Picket

FORD BATHERS

BOWDLER 1

Note;—
English ------
Natives ......
The figures 1. 2. 3. 4
denote the relative periods of
the fight and correspond to the
red figures on margin of
dispatch.

C. Heaphy.

# Trench warfare

ENVIRONS OF BEAUMONT-HAMEL AND SERRE, FRANCE, 1916

The First World War saw a greater use of mapping for military purposes than any previous conflict. On the Western Front alone, about 34 million official maps were printed for the use of the British army. These varied widely in their coverage and purpose, from large-scale trench maps designed as locational aids to smaller-scale overview maps intended to support high-level planning and administration. Standard-issue sheets included as much detail as possible about German trenches (which were normally overprinted in red) but far less about trenches dug by the British and their French allies (overprinted in blue). In many cases, special secret editions were prepared that showed the Allied trenches too.

This example comes from a set of around 30,000 maps brought together for the use of the researchers tasked with writing the British government's Official History of the Great War. It is made up of portions of several 1:10,000-scale map sheets trimmed and joined together to cover the area required – in this instance, part of the département of Somme in northern France. The resulting collage has also been heavily annotated in coloured pencil and ink, with an entirely hand-drawn patch glued over part of the right-hand side. These additions show the distribution of units within the British army's VIII Corps on 1 July 1916, the first day of the Battle of the Somme. To the south, the 29th Division have Beaumont-Hamel in their sights; to the north, the 31st Division stand ready to attack Serre; and the 4th Division are poised between the other two.

None of the ambitious Allied objectives for this attack (marked here as dark blue lines) were achieved. Despite making some initial gains, the British were forced to retreat as their opponents successfully defended their position. After 24 hours, the Allies had made no significant gains in this area and very few elsewhere on the Somme. Casualties had been exceptionally heavy: more than 19,000 British deaths and twice as many injuries. The Newfoundland Regiment (formed of men from the Dominion of Newfoundland) suffered a casualty rate of more than 90 per cent. Large numbers of dead and wounded men would continue to be a feature of the Battle of the Somme throughout its 141 days.

Alongside the cost to human life, the conflict wreaked terrible damage upon the physical environment. Just as the landscape detail of maps such as this one is obscured by overprinting and annotations, so was the peacetime world of roads, woods and villages obliterated by the network of trenches.

The Western Front during the First World War was far from being the first situation where dugout fieldworks were used in a long military campaign, but it was certainly the most complex and enduring example of trench warfare to date. In the popular imagination, it remains the archetype of combat by attrition, characterised by achingly slow progress and retreat, and extended periods of stalemate. For some, it is the ultimate illustration of the futility of war.

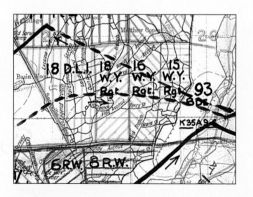

PALS' BATTALIONS: Some British army units were made up of friends, neighbours and workmates who joined up together, fought together and often died together. The men in these four battalions came from Durham, Bradford and Leeds.

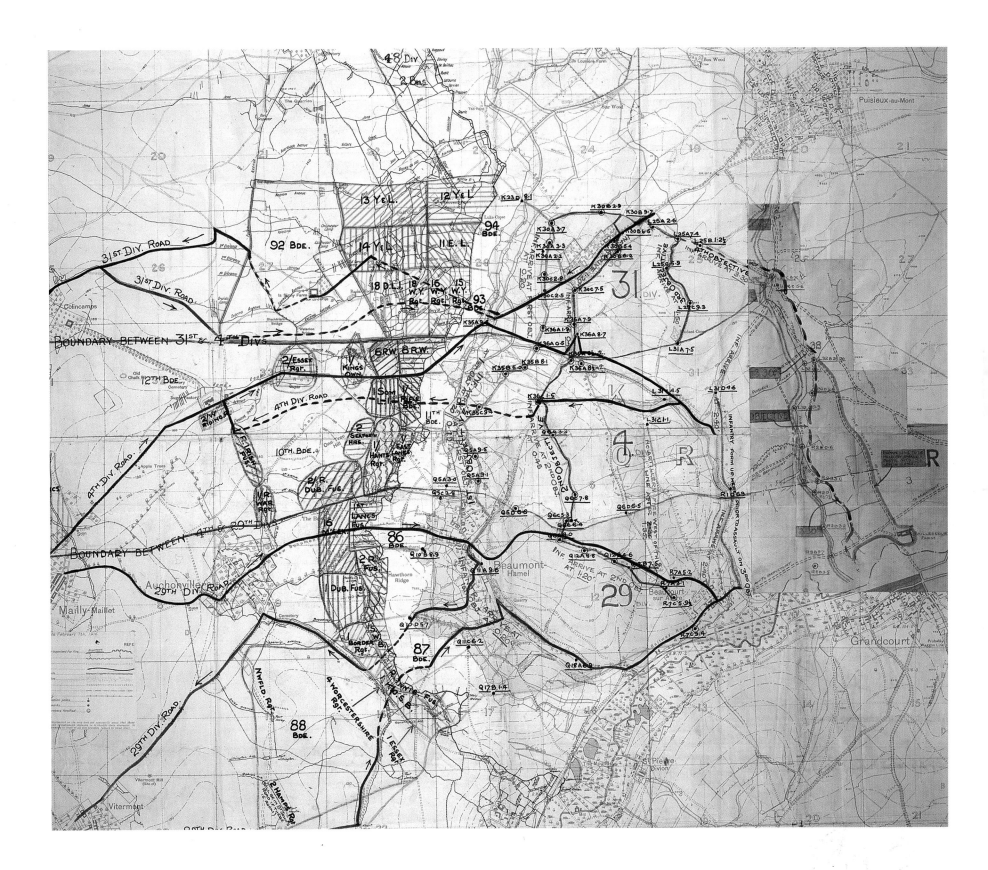

# Contours of conflict

This sketch map was made to illustrate the war diary of 2/18th Battalion, the London Regiment (also known as the London Irish Rifles) for May 1918. The archives contain many thousands of similar records compiled by army units fighting in the two world wars. They are not personal diaries but official journals of daily activity and progress, often supplemented by fuller accounts of combat and casualties. Small maps are frequently included to aid the explanations of events and enliven the written word. Some, like this one, are hand-drawn on scraps of paper; others are neatly cut from printed sheets and annotated.

At the beginning of May, the battalion was fighting in the small area covered by this map, which lies a few miles east of the River Jordan and some distance north of the Dead Sea. It was tasked with attacking Bulaybil (here spelt Bileibil, at upper right), near to the Ottoman stronghold of Shunat Nimrin (off the map, to lower right). The sketch marks the positions of the battalion's four companies during 1–2 May (in red) and 3–4 May (in blue). East is roughly at the top. The commentary in the war diary reveals that the battalion was subjected to intense fire from its opponents, often at close range. 22 men died and 108 were wounded.

The British would have found this landscape punishing even without constant enemy bombardment. The map's prominent, closely-spaced contour lines present a strong impression of rugged terrain. Nicknames such as Bucket Hill, Surprise Hill and Blackheath Ridge were devised to distinguish one otherwise nameless piece of high ground from another. Even relatively low hills could assume great importance as vantage points during military action. In this harsh environment, streams such as Wadi Nimrin (shown at right) could also be significant barriers to progress.

The battalion's activity was one component of a wider venture now known as the Second Action of As-Salt or the Second Battle of the Jordan. On 30 April, a combined force of British, Australian, New Zealand and Indian soldiers had crossed from the western side of the River Jordan to attack enemy-held territory to the east. Their aim was to gain control of this region – then known as Transjordan – by capturing the cities of As-Salt and Amman and their surroundings. This plan failed. Although As-Salt swiftly fell to the Allies, they were unable to take Shunat Nimrin. Ottoman and German counter-attacks forced them to give up the territory that they had gained, and by 5 May all of the surviving troops had retreated westward across the river.

Ultimately, however, Allied operations in the Middle East proved more successful. The Ottoman Empire, which had joined the war in the hope of regaining territory that she had previously lost to Russia, was defeated and broken up. Under the terms of the peace treaties, Transjordan was made a British protectorate. The region remained in British hands until 1946, when it gained independence as the Kingdom of Jordan.

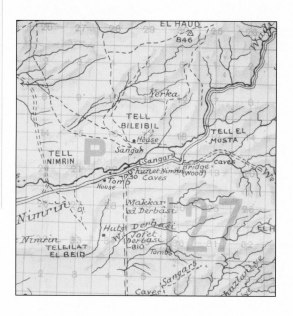

UNFAMILIAR TERRITORY: Unlike their counterparts on the Western Front, Allied forces fighting in Transjordan had no access to accurate and detailed printed mapping. Instead, they had to rely upon relatively small-scale maps, such as this one.

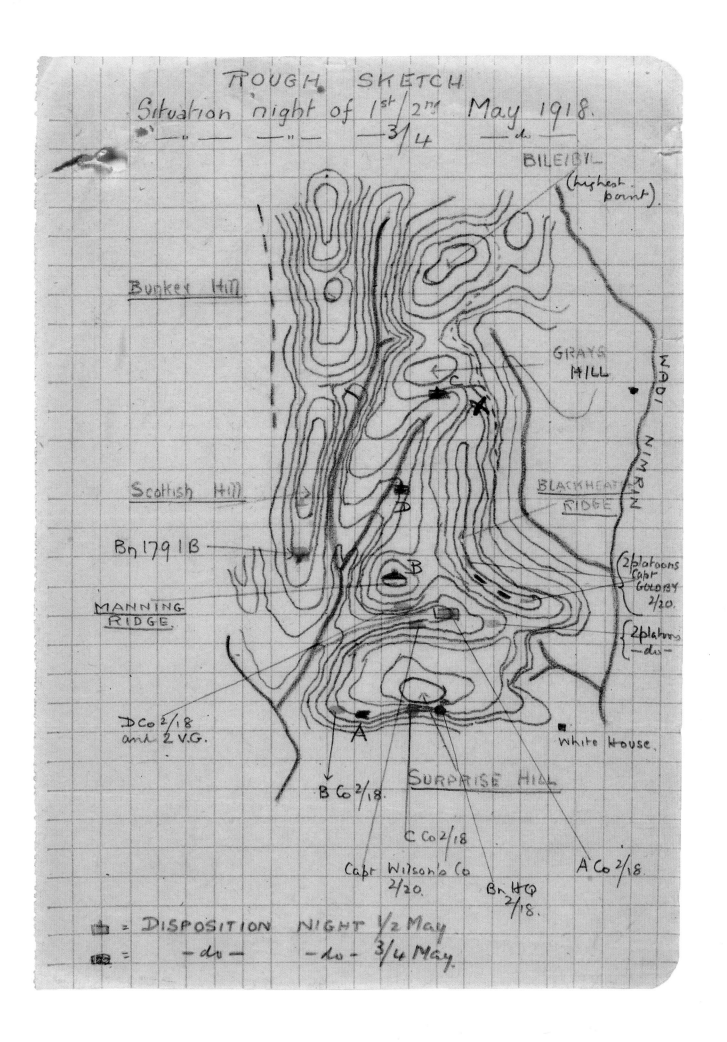

ROUGH SKETCH
Situation night of 1st/2nd May 1918.
— " — — " — 3/4 — do —

BILEIBIL
(highest point)

Bunker Hill

GRAYS HILL

WADI NIMRIN

Scottish Hill

BLACKHEATH RIDGE

Bn 179 I B

B

2 platoons Capt GOLDBY 2/20.

2 platoons — do —

MANNING RIDGE.

A

D Co 2/18 and 2 V.G.

White House.

B Co 2/18.

SURPRISE HILL

C Co 2/18

Capt Wilson's Co 2/20.

Bn HQ 2/18.

A Co 2/18.

⌷ = DISPOSITION NIGHT 1/2 May
⌷ = — do — — do — 3/4 May.

# 'The escapee's most important accessory'

The history of drawing and printing maps onto fabric can be dated back to ancient China. Although such maps have sometimes been made for decorative purposes (see page 235 for an example), lightweight fabrics also offer some practical advantages over paper, especially for military use. They are thin but durable – silk especially can be folded or crumpled repeatedly in a manner that would leave paper torn to shreds – and they can be handled noiselessly.

During the Second World War, about 250 different maps – some 1.75 million copies altogether – were printed on fabric for the use of the United Kingdom's armed forces. The impetus to create them came from MI9. This was the branch of the British intelligence services that was responsible for aiding attempts by military personnel to evade capture by the enemy or to escape once caught. The person credited with realising how useful silk maps would be in those circumstances was Christopher Clayton Hutton, an intelligence officer who insisted that the combination of map and compass was 'the escapee's most important accessory'. Believing (probably incorrectly) that no existing military suppliers could help him, Hutton approached the publisher John Bartholomew, who supplied him with paper maps to copy for his printing experiments.

Most escape and evasion maps were produced at relatively small scales – this example is at 1:2,000,000 – with breadth of coverage prioritised over fine detail. To begin with, the maps were printed in three or four colours, with pectin added to the ink to stop it running and to ensure a clear image. Here, the roads are in red, the all-important international boundaries in green, and other details in black. Not all 'silk' maps were actually made of silk, which was a scarce material in wartime; this example is printed on rayon.

Access to maps was thought to be especially valuable for airmen, who might survive being shot down over mainland Europe. A plan was devised to sew silk maps into the lining of air force uniforms. Such doctored clothing was to be identified by discreet labels bearing the word 'monkey'. Maps were also smuggled into prisoner of war camps, within items such as gramophone records and playing cards, to help men who were planning to escape. One clever scheme, devised with the co-operation of the board games manufacturer J C Waddington, involved hiding escape kits inside special versions of *Monopoly*.

Intelligence officers also developed many other practical aids and pieces of advice for men trying to escape from enemy hands or to avoid detection as British servicemen. These ranged from small tins of concentrated food to instructions for how to carry a bag like a French peasant. Despite this variety of gadgets and guidance, Hutton's opinion of the unique value of maps was apparently vindicated in practice. Of the 35,000 members of the Allied armed services who escaped from or evaded capture during the war, it is estimated that about half used a fabric map to help find their way home.

NEUTRAL COUNTRY: Allied personnel trying to escape from southern Germany often headed for Switzerland.

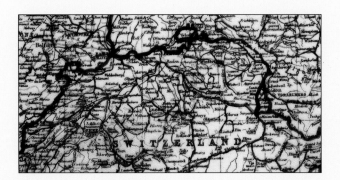

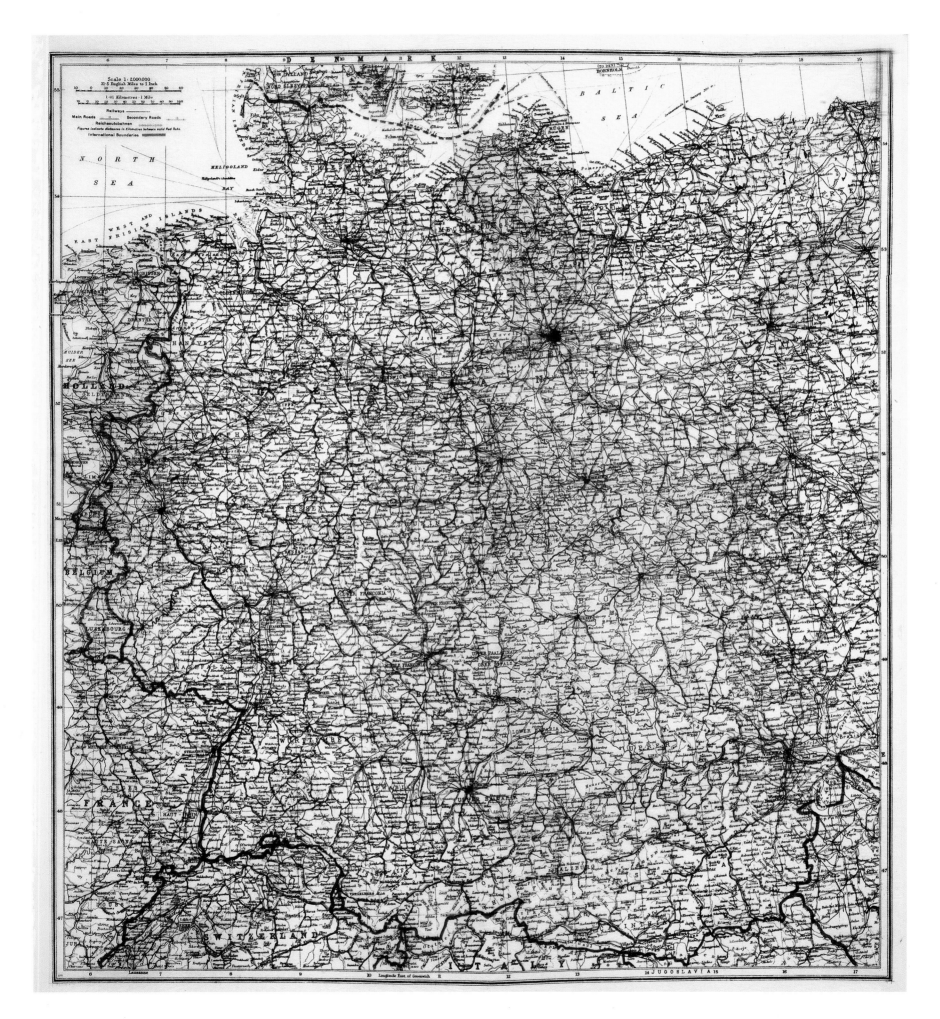

# Neptune comes ashore

The stark simplicity of this map belies the complexity of the military operation that lay behind it. This was the Allied plan to cross the English Channel and land in Normandy on D-Day, 6 June 1944. Codenamed Operation Neptune in an allusion to the Roman god of the sea, it formed the first step in the invasion of German-occupied France. The detail of the map is stripped back to a degree that allows the essential features of the venture to be seen clearly. Apparently extraneous pieces of information, such as place-names some distance from the focal area, serve the function of orienting the reader more quickly than a bare outline of the coast.

D-Day had been many months in the planning, and had from its inception been a co-operative enterprise between the United States and the British Empire. The coast of Normandy was chosen as the point of entry and divided between the Allies; the Americans were allocated the west and the British and Canadians the east. All of the arrangements were, of course, kept absolutely secret. An elaborate deception was devised to convince the German authorities that the Allies were planning to enter Pas-de-Calais, at the eastern end of the English Channel. This allowed the invaders to surprise the Germans when they landed. Targeted bombing over the preceding months had also severely damaged German defences in Normandy.

The invasion was inevitably an amphibious operation, requiring the combined effort of the Allied armies and navies. Five convoys of vessels departed from England on 5 June – a day later than originally planned, owing to bad weather conditions in the Channel – and reached Normandy early the following morning. Their routes are marked on the map as blue arrows, each labelled with the initial letter of the codename for its allotted landing beach. From left to right, these are: Utah, Omaha, Gold, Juno and Sword. The arriving Allied ships bombarded the coastal defences, causing sufficient damage to allow the disembarking troops to break through them. Additional support came from airborne assaults, shown here as pale green arrows. Arriving first, the airmen were able to gain control of key points, such as bridges, before the Germans were aware of the invasion from the sea.

The Allied forces gained less ground on this first day than they had hoped and lost about 7,000 men in the fighting. This map is an appendix to a report evaluating the level of opposition that they encountered on the landing beaches. Nevertheless, the operation was deemed a success and gave the Allies a small but secure foothold on the continent. The amount of territory under their control increased steadily over the succeeding weeks until by the end of August they had freed most of north-western France. Although another eleven months of fighting would follow D-Day before the final German surrender came, Operation Neptune provided the Allies with a solid base for beginning the liberation of occupied Europe.

FORMING THE PLAN: The seven most senior commanders involved in planning Operation Neptune are shown here seated in front of a map of Western Europe. The American General Dwight D Eisenhower is in the centre.

MAP SHOWING THE GENERAL LAYOUT OF THE AIR AND SEABORNE LANDINGS

N

BOULOGNE

ALLIED NAVAL COMMANDER EXPEDITIONARY FORCE

COMMANDER 21ST ARMY GROUP

FORCE L
7 ARMD DIV
(Follow Up)
Landing on
FORCE G BEACHES

Comd 1ST U S Army — NAVAL COMMANDER WESTERN TASK FORCE

NAVAL COMMANDER EASTERN TASK FORCE — Comd 2ND Brit Army

FORCE B
29 DIV
(Follow Up)
Landing on
FORCE O BEACHES

FORCE
U O G J S

NIGHT ROUTE

NIGHT ROUTE

Dieppe

PARACHUTE A/C ROUTE IN

PARACHUTE A/C OUT

GLIDER ROUTE IN AND OUT

DAY ROUTE

Fécamp

Pte de Barfleur

4 US 1 US 50 3 CDN 3 BR
DIV DIV DIV DIV DIV

CHERBOURG

LE HAVRE

ROUEN

DAY ROUTE

Cap de Carteret

Grandcamp

Port en Bessin

Arromanches

Cabourg

NIGHT ROUTE

Carentan

Isigny

Ouistreham

la Haye du Puits

BAYEUX

Tilly sur Seulles

CAEN

Troarn

Mondeville

Lisieux

ST LO

Coutances

Evreux

Falaise

Vire

Granville

Argentan

Dreux

St.Malo

Avranches

ST BRIEUC

Dinan

Scale 1:1,000,000

Miles 10 5 0       10      20      30      40      50 Miles

ALENCON

CHARTRES

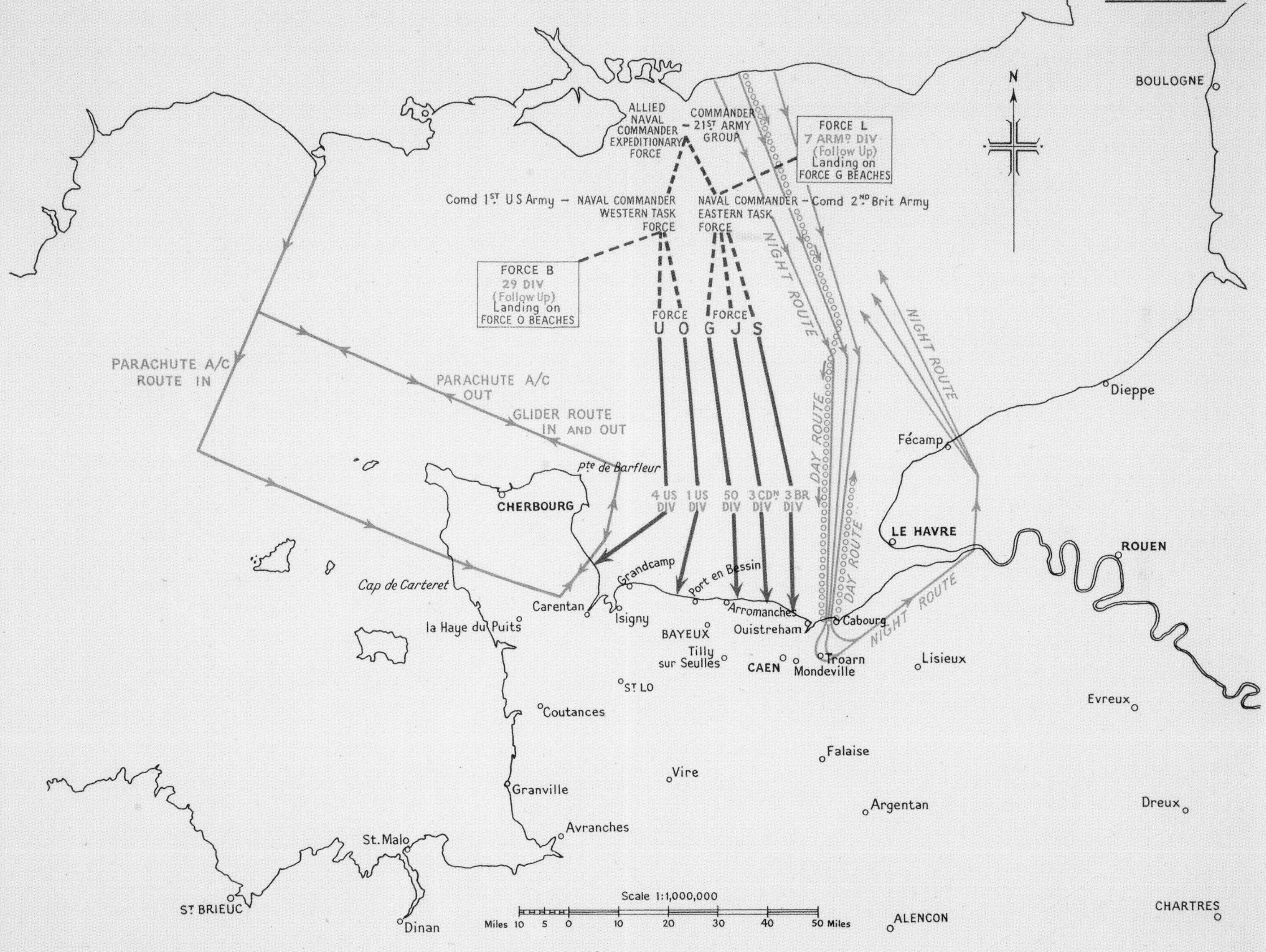

# A new type of weapon

NAGASAKI, JAPAN, 1945

After the Allies defeated Germany in May 1945, the Second World War ended in Europe but continued in eastern Asia and the western Pacific. Plans were drawn up to invade Japan – the only remaining Axis power – but this was expected to be extraordinarily costly to human life on both sides. In the hope of ending the war more quickly, the Allies decided instead to use the powerful nuclear weapons then under development in the United States. On 6 August a uranium bomb was exploded above Hiroshima, followed three days later by a plutonium bomb above Nagasaki. This was a controversial decision, much debated by historians ever since: were these bombings justifiable in the context of the war, and were they the decisive factor in Japan's surrender soon afterwards?

A party of British officials who visited Japan later that year was concerned not with those questions but with the physical effects of the new weapons upon the environment and upon human beings. Led by Professor W N Thomas, a civil engineer and expert in the impact of explosives, its remit was to gather information that would help the United Kingdom to prepare for any future nuclear war. This map of Nagasaki is one of four appended to its report. Based on a pre-existing military map, it was overprinted by the military mapmakers Geographical Section, General Staff with information about damage caused to the city. The concentric circles that mark distances from the centre of the blast zone are uncomfortably reminiscent of an archery or shooting target.

Nagasaki's history had been shaped by its geography. Its name means 'long cape' – a reference to the peninsula that extends southward from the area covered by this map – and its natural harbour made it an important port, and consequently a centre for shipbuilding. Surrounding mountains (depicted here with brown contour lines) had confined urban development to a fairly narrow valley. The direct impact of the bomb affected a relatively compact area north of the city centre, corresponding roughly to the inner three of the concentric circles. However, the fires that it caused spread much further, particularly within the areas shaded solid red. Many of the city's buildings were made chiefly of wood in the traditional Japanese style and were quickly destroyed.

Professor Thomas described Nagasaki in late 1945 as 'novel and eerie'. It looked less like a bomb site than 'a city struck by a tremendous hurricane' and 'an industrial slum'. By that time, at least 40,000 people had died as a result of the attack – some immediately and others later, from horrific injuries or radiation sickness – and another 60,000 were left seriously hurt. It was clear to the British officials that any future nuclear attack on the United Kingdom would cause utter devastation. The expected level of casualties would place exceptional strain upon medical and rescue services. The findings of their report were deemed so sensitive that they would remain secret for nearly 50 years before being declassified in 1993.

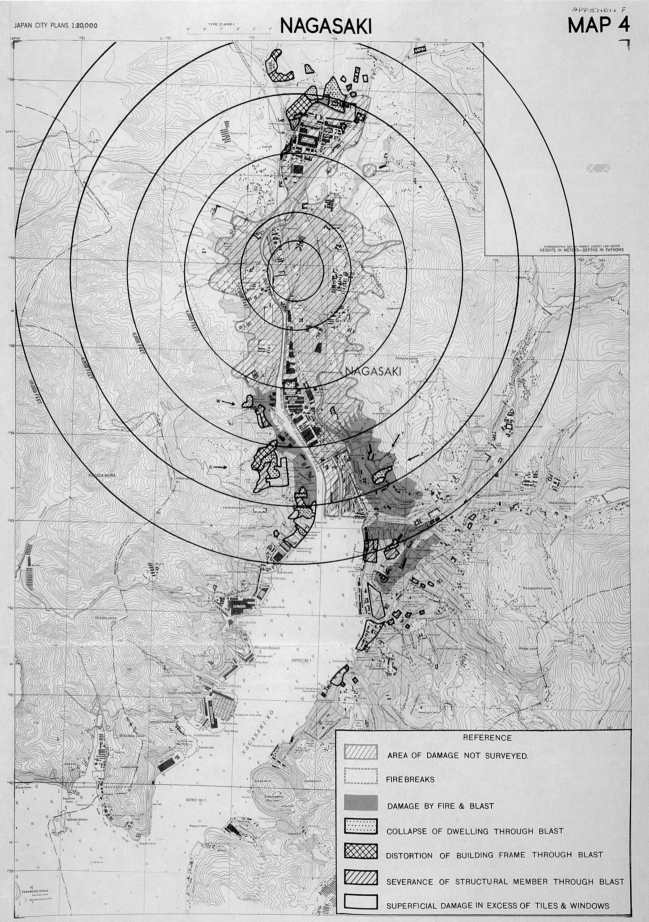

# NAGASAKI

JAPAN CITY PLANS 1:20,000

HYDROGRAPHIC DATUM NEARLY LOWEST LOW WATER
HEIGHTS IN METERS—DEPTHS IN FATHOMS

NAGASAKI

NAGASAKI KO

REFERENCE

AREA OF DAMAGE NOT SURVEYED.

FIRE BREAKS

DAMAGE BY FIRE & BLAST

COLLAPSE OF DWELLING THROUGH BLAST

DISTORTION OF BUILDING FRAME THROUGH BLAST

SEVERANCE OF STRUCTURAL MEMBER THROUGH BLAST

SUPERFICIAL DAMAGE IN EXCESS OF TILES & WINDOWS

O.R. 4271

Overprints Drawn at G.S.G.S.(A.M.) 1946

# CHAPTER FIVE

# CHARTING the seas

## '... THOU AT HOME, WITHOUT ... TIDE OR GALE, CANST IN THY MAP SECURELY SAIL ...'

Robert Herrick's verse evokes the delights afforded to armchair travellers by sea charts, which invite us to wander the waters of the globe, immune to the dangers and discomforts experienced by real-life sailors. In this chapter we sail across time and the oceans, to explore a range of possible things that can happen at sea – not all of them for the worse, although shipwreck, mutiny and pirates feature in these pages.

The British Isles are surrounded by seas, and a natural interest in sailing is reflected in the many charts in these archives. From earliest times, mariners carried maps in their minds, and from around the 14th century they began to commit these to parchment. A portolan chart from this era on page 139 shows the Mediterranean, crossed by navigational lines that were of practical help to sailors. The idea of capturing information for use on future voyages is also evident in the sea charts and coastal views with which naval officers illustrated their ships' logs, from the late 17th century. An early example on page 147 is by

Grenvill Collins, whose journals' pages are crowded with drawings.

From the 16th century, charts were used in England to inform trade and defence policy. Examples here show the defensive capabilities of three ports: Plymouth (opposite), Tangier while it was briefly in British hands (page 143), and the Russian stronghold of Kronstadt (page 163). Other charts record battles at sea. Page 141 features an Elizabethan chart of Cadiz which shows how English ships made a daring raid upon Spain under Sir Francis Drake, while Nelson's victory at the Battle of the Nile is the focus on page 159. Ships also conveyed troops overseas; a chart showing General Wolfe's voyage up the St Lawrence River to action at Quebec is on page 153.

Not all of these charts are of English provenance, reflecting the often cut-and-thrust nature of life on the high seas, in times when the line between pirate and prize-taker was rather fine. In the mid-18th century during the Seven Years' War, British captains captured many foreign ships as prizes

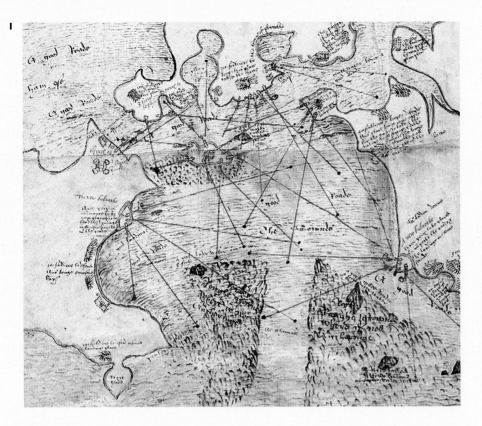

**1** THE LIE OF THE SEA: This manuscript chart was made about 1587 to show how best to defend Plymouth, in expectation of an invasion by the Spanish Armada. Any ship approaching the Hoe would be covered by arcs of fire from cannon and bands of militia stationed around the Sound. In the event, Drake sailed out to engage the Spanish offshore.

**2** PLAIN SAILING: Decoration from an 18th-century naval officer's hand-written manual on mathematics.

**3** CHART BY NELSON: Horatio Nelson, while a more junior officer, proved himself a competent chart-maker, as we see in part of a manuscript survey of St John's in the Virgin Islands, dated 1784.

– along with their charts; an example is on page 151. Other dangers abounded at sea, too. It was a shipwreck off the Isles of Scilly (see page 149) that sparked the Longitude Prize.

Official British sea charts were not printed until the 19th century. Before then, ships' captains had to buy their own, and commercial map publishers vied to provide charts for mariners, as well as atlases for gentlemen's libraries. A chart of the Arctic from one of the latter is shown on page 161, which has much decorative detail, but little practical value to the seafarer.

The Admiralty sponsored voyages of exploration, and charts for many of these great journeys are in the archives. Chart-making was part of a British naval officer's training, refined by long years at sea, and in some notable cases by a lineage of survey experience. James Cook's voyages in the Pacific (1768–1779) are the most famous of these (see page 155); Cook trained George Vancouver, who made an epic survey of America's north-west coast, and William Bligh, who charted from the *Bounty* and

the *Providence* (see page 157). On Bligh's successful breadfruit voyage Matthew Flinders served as midshipman, learning skills useful for his survey of the Australian coast in 1801–1803. Flinders trained his nephew John Franklin on that circumnavigation. Franklin was a friend of William Parry (see page 161), and they both searched for a North-West Passage in the frozen Arctic in the 1820s.

From 1800, the newly-established Hydrographic Office published Admiralty charts, aimed to be authoritative and up-to-date navigation aids for the Royal Navy and other mariners. Fewer manuscript charts were made after this, as navigators simply added hand-written information about the business at hand to their printed charts. From medieval portolan to pre-war printed Admiralty chart – six centuries of seafaring are represented here. While some of these charts stayed at home, many arrived here from afar, survivors of the rigours of life at sea. They bear witness to long, tedious hours in fog and becalmed waters, subject to tides and gales, and to the skill and adventures of their makers.

**4** SURVEY IN SOUTH AUSTRALIA: This coastal view and drawing of a survey boat at 'Yanky-lilly' (modern Yankalilla near Adelaide) evokes early exploration of this area in 1836, and the method of making coastal charts more generally.

**5**

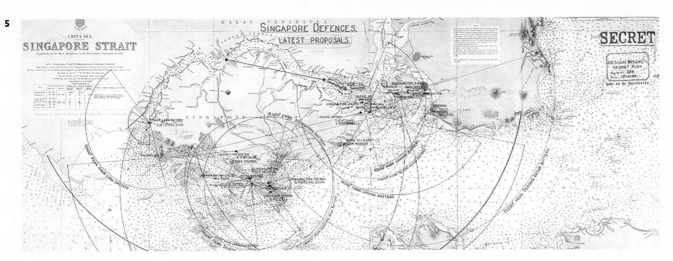

**5** SECRECY IN SINGAPORE: A printed Admiralty Chart of Singapore Strait is marked 'secret – not to be duplicated' because it bears additions in red to show proposals for defences in 1936. An assumption that attack was not expected from the north would prove to be a grave tactical error in 1942.

**6** HEDGEHOGS ON THE BEACH: This top secret 'Bigot' chart-map with coastal profile shows defences faced by Allied troops on Omaha Beach in the Normandy D-Day landings of June 1944. A note about 'hedgehogs 3 to 4 deep' in a random pattern refers to spiked metal obstacles in the water.

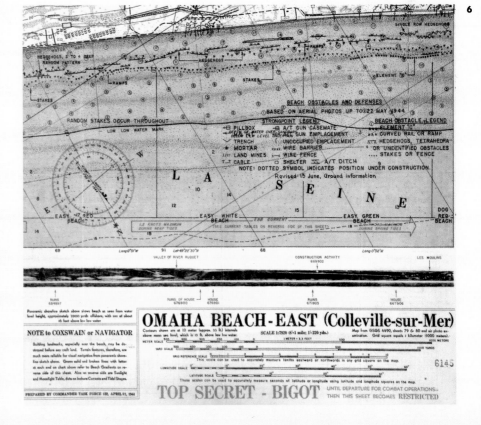

# A medieval mariner's compass

EASTERN MEDITERRANEAN AND BLACK SEA, 14TH CENTURY

The sea spray had already long evaporated from this portolan chart, when it embarked on a second life on land in the 16th century. While decorative sea charts for display at European courts are found in many map collections, often with still-bright colours and gilding, this is a rare survival of a working chart. It would have been taken to sea to assist with navigation, then discarded when it became obsolete. Parchment was valuable, however, and this spared the chart from destruction. It went on to serve as the cover of a Tudor Exchequer customs account which lists payments made in 1546; in this phase of its existence it acquired a clerk's scrawl across its surface. It was rediscovered in 1953 by a conservator in the course of repairing the account.

Portolan charts were made by medieval European mariners who plotted coastlines along constant compass bearings. Drawn here in red ink, when put on the chart these bearings are known as rhumb lines. They are often shown radiating from compass roses or compass stars, which represent the 32 directions of the mariner's compass. There are four compass roses on this chart, along the top edge.

Only part of this chart appears to have survived. Complete portolans always extended to the eastern end of the Black Sea, but this section covers just the western Black Sea and the eastern Mediterranean. It is possible that the chart originally continued further westward, too, perhaps to Gibraltar, or even into the Atlantic. Whatever its original size, this surviving portion still shows a wide area, and compares quite well for accuracy with much later charts. Between the east coast of Greece at the lower edge and the west coast of Turkey at the top lie the shores of Bulgaria and Romania, the Crimea, and the islands of Crete and Cyprus. The outline of the coasts was drawn in black ink, now faded to brown, while some islands in the Aegean Sea are coloured red.

The chart has been dated tentatively to the early 14th century, on the evidence of the many place names, in both red and black ink, although they are very hard to read. They all lie along the coasts and not in the interior, which was a constant feature of sea charts through the centuries. The place names are in Italian, with a particular spelling which may point to a Venetian origin. At this time, Venice was a flourishing maritime republic, with trading connections across Asia, including those forged by its famous citizen Marco Polo. It was also one of the centres of portolan production; the functional style of its charts contrasted with those of the more decorative Catalan school.

We can only guess which of these seas this chart sailed, and in which ports it found safe harbour around seven centuries ago. Its original purpose was hidden, and only 60 years ago did it emerge again, to take its place as probably the oldest map in the archives.

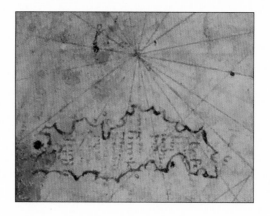

CYPRUS: This detail from the top right hand corner of the chart shows the island, a centre for trade with the Orient.

# Confined to cabin

CADIZ, SPAIN, 1587

This chart shows Sir Francis Drake's 'singeing of the King of Spain's beard' – a daring attack by the English fleet on the nascent Spanish Armada forming in Cadiz harbour. It was drawn, however, not to celebrate this audacious naval exploit, but to explain and justify the mapmaker's part in it. William Borough, as second in command, made plain to Drake his views that the raid – carried out when some of the fleet were still to arrive from England – was rash and not in accordance with naval procedure. Drake 'tooke it in very ill parte', as Borough reports it. He removed his forthright vice-admiral from command of his ship, the *Golden Lion*, confined him to his cabin, and sought to have him tried for mutiny. Cooped up below decks, Borough drew this chart and sent it to the Lord High Admiral with a letter in which he gave his version of events.

Borough's chart tracks successive positions of the main English galleons during the raid on Cadiz, the major port on Spain's south coast. Letters in black denote 'first' and later locations for the *Lion*, Drake's flagship the *Bonaventure*, and the rest of the fleet. Borough claims his share of the action in a note labelling galleys 'dreven back by ye Lyon'. The chart makes apparent the dangers to which he felt that his ship was exposed. Specific Spanish threats, including galleys poised to attack the *Lion*, are listed in red at left. The chart also shows the town's strong defences, which had caused Borough to urge caution in approaching it. As Master of the Queen's Ships, he had good reason to frown upon the risks which Drake favoured.

This was not the first chart made by the 50-year-old Borough, who was a good example of that Elizabethan genre, the man of many parts. In his youth he had served on Richard Chancellor's expedition to search for a North-East Passage to Cathay (China), organised by Cabot. He spent many years establishing trade routes to Russia as a captain for the Muscovy Company, and held posts with the Cathay Adventurers and the Levant Company. In 1580 he entered Crown service as a naval administrator, and became Comptroller of the Navy. This was not just a desk post – he also went to sea to capture pirate ships, and sailed with Sir John Hawkins to the Azores in 1586 in the *Golden Lion*. Borough made maps in all these roles; he charted Russian waters as a merchant, and drew harbour works at Dover and Rye as Comptroller, while this battle plan was drawn during his service as a naval officer.

Despite Drake's displeasure with his subordinate's caution, this was not the end of Borough's naval career. The eloquence of this chart may well have persuaded Borough's fellow Navy Board members that he had good reasons for his actions. Acquitted of Drake's charges, he served against the Armada in the following year, and his name is remembered today as a famous navigator and explorer.

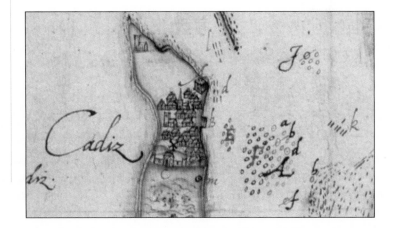

ACTION AT CADIZ: The town, its buildings drawn in perspective, sat on a promontory overlooking the harbour. It was defended by several forts, and guns upon the town walls and along the harbour side, one of which at 'm' scored a hit on the *Lion* (at 'f'), injuring a sailor.

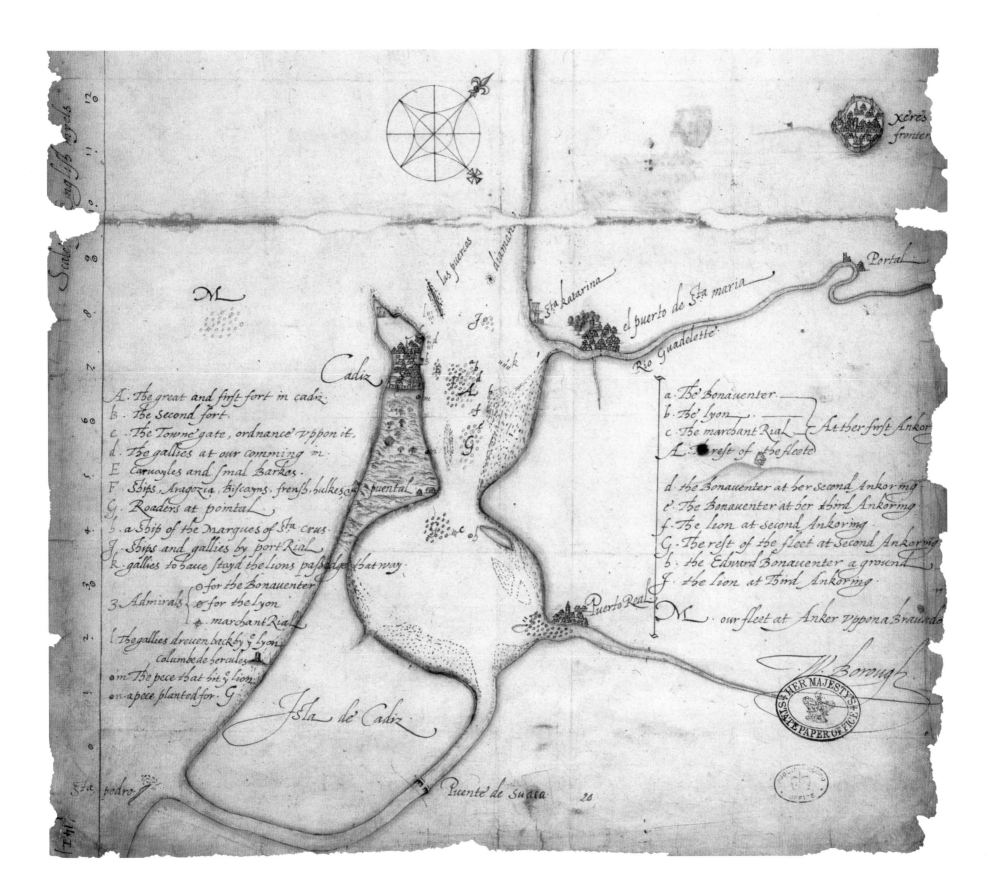

Scale Englifsh myls

Cadiz

Isla de Cadiz

Sta pedro

Puente de Suaca   20

Las puertas   diamante

Sta Katarina

el puerto de Sta maria

Rio Guadelette

Portal

xeres fronter

Puerto Real

puental

A. the great and first fort in cadiz
b. the Second fort
c. The Towne gate, ordnance vppon it
d. The gallies at our comming in
E. Carvoyles and smal Barkes
F. Ships, Aragozia, Biscayns, frensh, bulkes
G. Roaders at pointal
h. a Ship of the Marques of Sta crus
J. Ships and gallies by port Rial
k. gallies to haue stayd the lions passedge that way

3 Admirals { o for the Bonauenter
              e for the Lyon
              b marchant Rial }

l. the gallies dreuen back by ye Lyon
   columbe de hercules
m. The pece that hit ye lion
n. apece planted for G

a. The Bonauenter
b. The Lyon
c. the marchant Rial } At ther first Ankor

A. the rest of the fleete

d. the Bonauenter at her second Ankoring
e. the Bonauenter at her third Ankoring
f. The lion at second Ankoring
G. The rest of the fleet at Second Ankoring
h. the Edward Bonauenter a ground
J. the lion at third Ankoring

M. our fleet at Anker vppon a Brauado

W. Borough

HER MAJESTY'S STATE PAPER OFFICE

# The dowry of a queen TANGIER, 1675

Located on the north-western tip of Africa, Tangier has long been a thriving city and major port. From the late 15th century it was controlled by Portugal and Spain, but came to the English in 1661 when King Charles II married the Portuguese Princess Catherine of Braganza. Along with Bombay in India, Tangier formed part of her dowry. The town was a strategic possession for far-off England, since it commanded the entry to the Mediterranean, and offered the potential to be a lynchpin of English naval strategy.

At the point when this plan was made in late 1675, a decade of work had been undertaken to fortify the town and make it safe for English ships. The outer ring of redoubts and forts to protect the harbour are shown complete. The large building at the back of the harbour was the newly-built Customs House, with adjoining weigh-house, warehouses and wharves. This is one of over 50 plans which record work on Tangier and its defences; these plans were sent to London, from whence finance and supplies were sought to maintain and develop the colony.

Tangier is here presented as an English stronghold, with the royal arms at top right. The bird's-eye view provides a sense of perspective, while pictorial elements vividly convey the scene. All kinds of shipping lie safely in the harbour, and the city is shown with the amenities to become a major trading port. The walls seem strong, with strategically-placed watchtowers, and an island fort at left. The handsome galleon at the harbour entrance appears illuminated by the rays of the splendid compass rose. All of these suggest prestige and power.

Yet in 1676, the year after the chart was made, a survey revealed Tangier to be effectively a garrison town, in that most of the population were military men and their families. There were no signs that it was set to become a civilian colony. This was despite the fact that, to try to encourage settlement, in 1668 the town had been given a charter equivalent to those of cities in England, with a mayor and corporation to run it, instead of the army. In the light of this, the blank space at the back of the plan where the town lay takes on a symbolic significance.

In reality, Tangier proved to be too expensive to maintain from a distance, with no other English colonies nearby. It was under near-constant attack from local Berber people, while Barbary pirates harassed merchant shipping in passage to and from the port. English rule ended in early 1684, with the entire population evacuated. Samuel Pepys, as Secretary to the Navy Board, went out to Tangier to oversee the demolition of the fortifications built over the previous two decades. The mole, which had only been finished in 1676 at a cost of £340,000, was mined, and all of these expensively-erected defences were blown up, to leave the town in ruins for the Sultan of Morocco's forces, who then gained control. This early, short-lived outpost of the British Empire thus ended in gunpowder and dust.

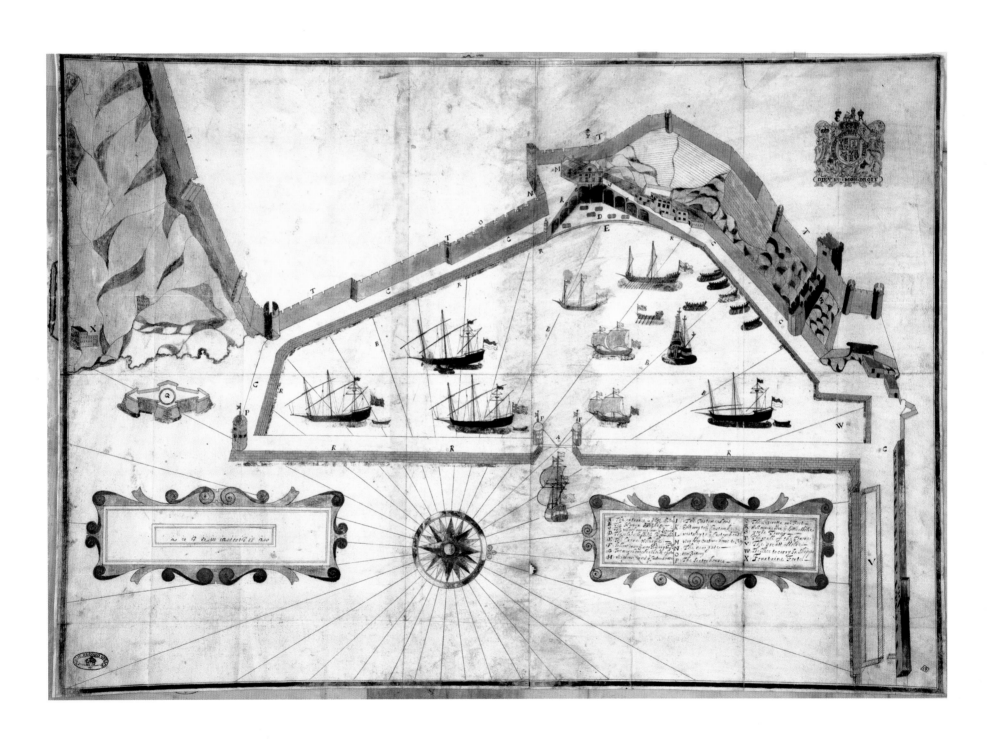

# Seas of the midnight sun ARCTIC OCEAN, c.1675

The icy waters portrayed on this 'Chart of the Sea Coasts of Russia Lapland Finmarke Nova Zemla and Greenland' lie chiefly within the Arctic Circle, where the sun shines day and night in the middle of the short summer. This area is to the extreme north of Europe. What is marked as Greenland at top left is really its eastern neighbour, Spitsbergen. Below lie the northern shores of Scandinavia. 'Finmarke' is part of modern Norway, 'Corelia' (Karelia) lies to right of the prancing goat, while Lapland forms the 'nose' of the horse's head shape. Below Lapland lies the White Sea, and the coastline then sweeps to the right across the top of Russia.

These were the shores past which William Borough (see page 140) and his brother Stephen had sailed in service with the Muscovy Company in the 16th century. Between the islands of Novaya Zemlya (at right) and the Russian mainland below it lies a narrow channel marked 'Fretum Burrough' on this map. This refers to the brothers' discovery in 1556 of what William called the 'Straits of Viagatz' (now the Kara Strait), which leads to the 'Tartarian Sea' and the Far East. At the centre of the chart's lower edge lies Archangel, which the Boroughs reached, enabling them to find an overland route to Moscow, to begin trade negotiations with Tsar Ivan the Terrible.

This chart was not intended for navigation. It lacks information useful to mariners, such as soundings and currents; warnings of shoals, rocks and sandbanks; and the locations of harbours. The space is filled instead with decorative details. Walruses eye a blowing whale at top centre, below a boat where a hunter stands with poised harpoon. Three people bearing baskets and a bow and arrows tramp through the Scandinavian interior. The ships, compass roses, and ornate cartouches were designed to appeal to an audience wealthier than seamen. The chart is a plate in the *Atlas Maritimus* of John Seller, one of the first Englishmen to challenge the 17th-century Dutch dominance of an expanding market for charts among the nobility and gentry of Europe. The first edition of the Atlas was published in 1675; our copy, which formerly belonged to the Foreign Office, may come from a slightly later edition.

In his keenness to issue attractive maps quickly, Seller drew heavily upon Dutch work. Many charts in his Atlas were made either by direct copying or by updating old copperplates which he had acquired in Holland; both were common practices among his contemporaries. This rare chart, however, was specially engraved by Stephen Board for Seller's English Pilot published in 1671. The more famous Atlas Maritimus was issued on a bespoke basis, with specific sheets such as this one added to meet customer requests. Another special service offered to buyers was hand-tinting, which made each plate unique. In this case, we like to think that the choice of colouring produced what may be some of the earliest images of Father Christmas.

SANTA'S FIRST OUTING? This figure crosses the snowy wastes of Novaya Zemlya in a sleigh pulled by reindeer, perhaps en route to Lapland or the North Pole. Another image shows him dressed in red and white, leaning on the scale bar as if chatting with the polar bear.

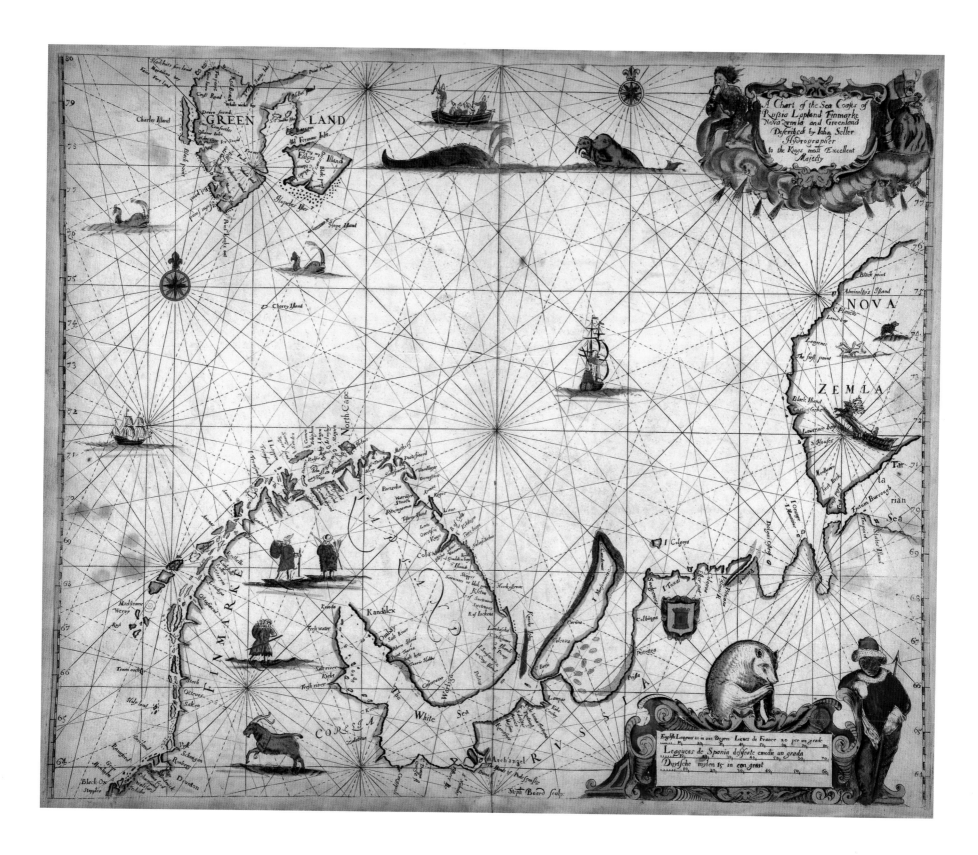

A Chart of the Sea Coasts of
Russia Lapland Finmarke
Nova Zemla and Greenland
Described by John Seller
Hydrographer
to the Kings most Excellent
Majesty

GREENLAND

NOVA
ZEMLA

Tartarian Sea

FINMARKE

CORELIA

White Sea

RUSSIA

North Cape

Archangel

English Leagues 20 in an Degree Lieues de France 20 per un grade
Legguas de Spania desseate cmedie un grada
Duytsche mylen 15 in een graat

# 'Where ye Fire & Smoake cometh out'

Captain Grenvill Collins was a 17th-century naval officer and hydrographer (also spelled Grenville, Greenvill or even Greenvile). His early career took him first to the South Seas (1669-1671) and then to the Arctic in search of a north-west passage to China and the East Indies in 1676, when his ship was wrecked but the crew were rescued. He was then sent on survey work in the Mediterranean from late in 1676 to 1679, as master of a succession of Royal Navy frigates (the *Charles*, *James*, *Newcastle*, *Plymouth* and *Lark*). Collins' handwritten log provides a lively account of these ships' travels, recording the dates they visited places, detail of weather, encounters with pirates, and instructions on how best to navigate into ports.

Collins' log includes sea charts and coastal views that show how places looked from onboard ship. He drew Falmouth, Alicante, Gibraltar, Malaga, Majorca, Tangier and the Bay of Naples, shown here, where the *Newcastle* arrived on 16 March 1678. As is typical of a chart, only the sea and major places along the shores of the bay are shown, including the island of Capri to the right. Useful information for sailors is given, such as the location of rocks and places of safe anchorage, and the numbers in the water are soundings, giving the water depth. A compass rose radiates rhumb lines for navigation, and the feathers point to north, slightly to the left. The scale is presented on a ribbon.

The chart shows Vesuvius billowing smoke, at the centre of the Bay of Naples. Dormant since the 13th century, it had last erupted in 1660, and was still active at the time of Collins' visit. He drew Vesuvius in plan to illustrate the written account of his intrepid ascent of the volcano, first by mule until it got too steep, then he climbed up on foot to the lip of the crater.

Collins's hydrographic skills are evident from the charts in his log, and when he returned to England in late 1679 he was commissioned to survey home shores, which he did for the rest of his life. *Great Britain's Coasting Pilot*, published in 1693, was the work for which he is best known, a beacon in the history of navigation in British coastal waters, and its level of detail doubtless helped to save many ships and sailors from wreck.

What happened to Grenvill Collins' log? A note on the front cover reveals that it is quite by chance that this rare survival of an early captain's log ended up in the Admiralty records. It reads 'This book bought off a stall in Moorfields the year 1774 for 6 shillings, and presented to Rear Admiral Man. Left by him to Captain Robert Man', who gave it to the Admiralty in 1788. Whatever tales the log might tell of its travels in the hundred years after it was written, it gives a fascinating insight into the man and his view of the world from on board a Royal Navy ship.

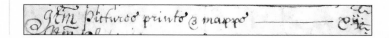

ALL THAT HE POSSESSED: An inventory taken at Grenvill Collins' house in Deptford (convenient for the dockyard) after he had died in March 1694 lists his worldly goods, including 'Pictures, prints & mappes' valued at twelve pounds. They graced the Great Dining Room, among furniture, carpets, plate, a looking-glass, pistols and carbines [short muskets]. Was Grenvill Collins' ships' log perhaps one of the 'books' listed with Collins' mathematical instruments – probably for survey work on the Coasting Pilot – kept with a 'close stoole' in a 'dark closet'?

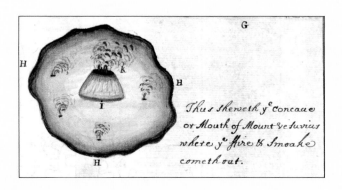

Thus sheweth ye concaue or Mouth of Mount Vesuvius where ye fire & Smoake cometh out.

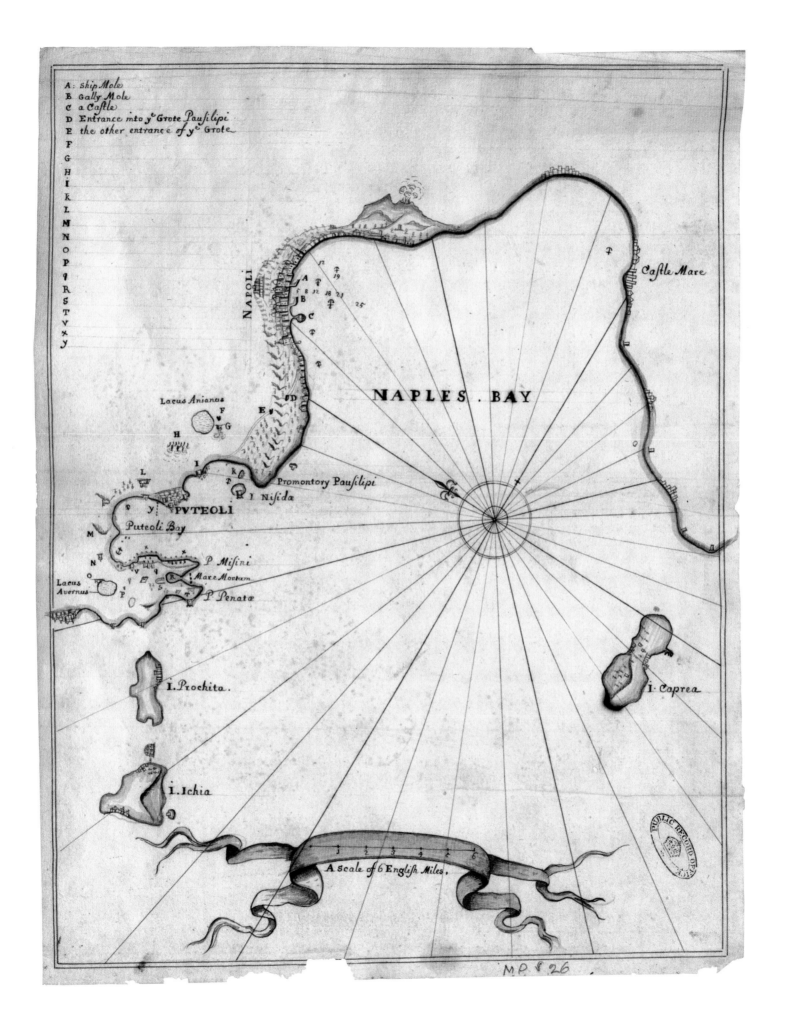

A: ship Mole
B: Gally Mole
C: a Castle
D: Entrance into ye Grote Pausilipi
E: the other entrance of ye Grote
F
G
H
I
K
L
M
N
O
P
Q
R
S
T
V
X
Y

NAPOLI

Castle Mare

NAPLES . BAY

A
B
C
D
E

Lacus Anianus
F
G
H

I
k
R
I Nisidæ

Promontory Pausilipi

L
Y
PVTEOLI
Puteoli Bay
M
N
O
P
Lacus Avernus

P. Misini
Mare Mortum
P. Penatæ

I. Prochita.

I. Coprea

I. Ichia

A Scale of 6 English Miles.

MP. 8 26

# A shocking shipwreck

How do you know where you are at sea? The consequences of error are evident on this chart, which, despite the fantastic array of creatures in its margins, has a serious purpose. It shows the scene of one of the greatest maritime disasters in British history when, on 22 October 1707, the Royal Navy lost four ships and nearly 2,000 men, in just one incident. Although Britain was at war with France, these losses were not due to enemy action. They happened because the sailors wrongly thought that they were in safe waters, and had no way to check their course.

The fleet, commanded by Sir Cloudesley Shovell, had been harrying the French in the Mediterranean, and was returning to winter in England. Heading for Plymouth, they encountered stormy weather which sent them off course, further westward than they realised. Instead of the familiar home waters of the English Channel, deep enough for their ships, in the dark and the rain they strayed into danger around the Isles of Scilly. Lookouts suddenly saw rocks and the loom of the St Agnes light. Shovell's flagship, the *Association*, and HMS *Romney* struck the Gilstone Rocks, while the Eagle went down off Tearing Ledge. All hands but one were lost. The badly-holed *Firebrand* sank later, with only 23 survivors. The other 14 ships heard warning shots and managed to escape.

This striking circular chart must have been drawn just after this tragic event. The scene of action was in the lower left, and the soundings reveal the shallowness of the inter-island waters, which combined with many rocks and shoals to make this a perilous place for ships to enter unawares. Nothing is known about the mapmaker, Edmund Gostelo, but the detail of his chart suggests that he was a native of Scilly. He includes all the islands, the sounds and roads between them, and buildings such as Tresco Abbey, churches and houses.

Figures in the chart relate to the incident. The warning lighthouse is pictured at lower right, together with a mermaid, whose song in legend lures sailors to destruction. The lower cartouche bears two mourning figures. By contrast, the border around the chart is covered with sketches of farming scenes and figures, as if copied from a pattern book simply to fill the space. Native and exotic animals mix in a bizarre tableau, where lions lie down next to rabbits (at right) while an elephant trumpets next to a dog (top left).

The chart may have been commissioned by the islands' governor, Sidney Godolphin, and sent to London to communicate details of this tragedy. News of the wrecks and loss of so many men, including a famous rear-admiral, shocked court circles and the public, especially since the fleet was so close to home. A lack of accurate means to calculate longitude – how far they had gone east or west – was thought to have been a contributing factor, which sparked calls to make seafaring safer and led to the offer of the Longitude Prize.

WRECKING ROCKS: The chart shows 'ye Exact Places where The Association [and] Rumny ... was Lost' by pictures of the ships overcome by water. It records above the *Association* 'On this Gilston Sir Clodesly was Lost'.

# Coffee with sugar: a captured French captain's map

This map of the French sugar island of Martinique was found among prize papers of the High Court of Admiralty, where it had lain since the capture of the French merchant ship on which it was being used during the Seven Years' War (1756–1763). *Le Constant* had been on its return journey to its home port of Marseille from Martinique laden with a cargo of sugar, coffee and cocoa, when it was seized somewhere in the vicinity of the British island of Anguilla, north of Martinique, in the summer of 1757. British naval captains could claim prize money by presenting the papers of a captured enemy ship to the court. *Le Constant* was pronounced a legitimate prize of war at the Vice-Admiralty Court on Anguilla in October 1757, and the two hessian sacks of the ship's papers were kept as evidence.

This manuscript map gives a general view of the interior of Martinique with its hilly terrain, trees, settlements and roads. It shows detail around the coast with ports, harbours, headlands and islands. Below the title in its decorative cartouche is a note 'Lauteur de lisle' which suggests that the map was perhaps copied from one by the important early 18th century French cartographer Guillaume Delisle. His original manuscript map of Martinique held at the Bibliothèque Nationale in Paris has the title, scale bar and compass rose in the same parts of the map, with similar detail of coast and hills, and using the same place names and anchorage symbol locations. Whatever its source, this manuscript map is evidence of the kind of working map used on board a French merchant ship, which by the nature of paper and the rigours of sea life have rarely survived.

The working papers among which the map was found cast light on shipboard life in the mid 18th century. They include a crew list,

table of cargo, ship's log, a list of suppliers and account books. There are many papers of the captain, Henri Balthazard Duprat, which form a vivid snapshot of his life at the moment his ship was seized. These include a handwritten copy of his birth certificate, a number of hand-drawn playing cards tucked in the cover of his notebook, and even a sheet of blotting paper. Promissory notes (notes of debits and credits) include one recording the loan of 'a negro cook' when his own ship's cook fell sick. Duprat's pocketbook contains notes about how to use marine charts.

The annual pilot's licence issued by Marseille port authorities for a year from 1756 notes that Duprat was then aged 32, and unmarried. His notebook contains drafts of his letters. To his lady friend he expressed his deep affection, saying that not a day passed without him thinking of her and he then goes on to mention marriage. To his brother in Marseille he wrote that '*ces voyages sont tres penibles et dangereux*' – and it had certainly proved a hard and dangerous voyage for him.

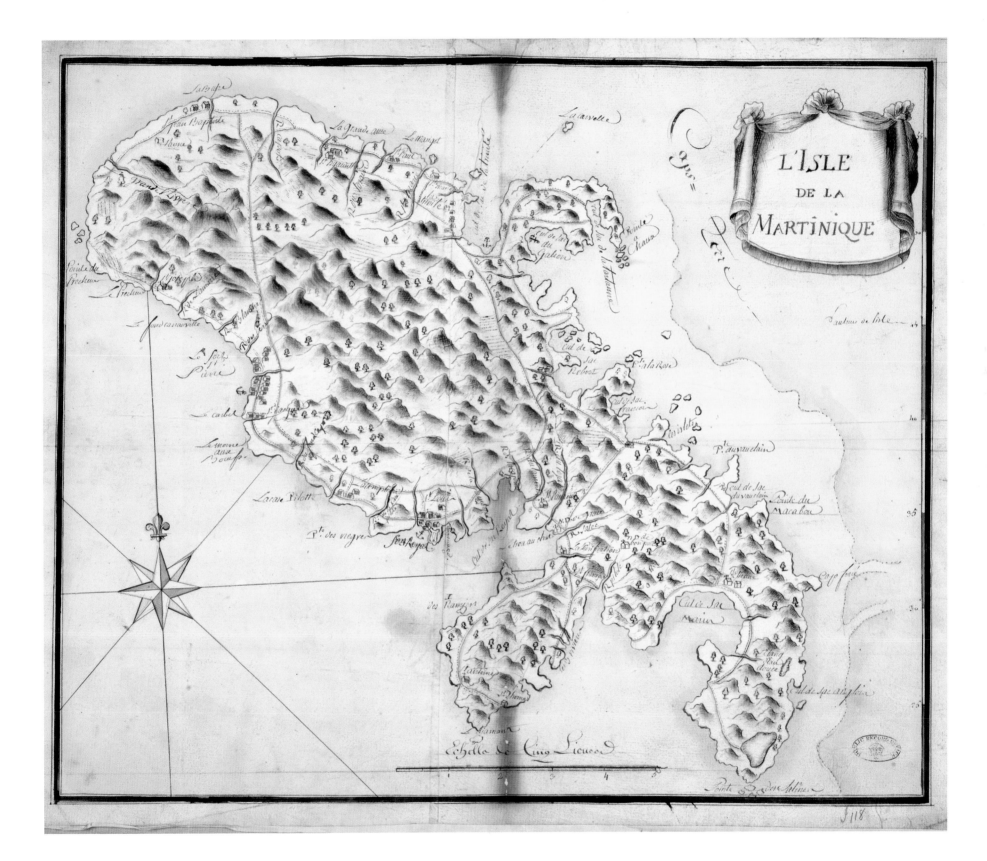

L'ISLE
DE LA
MARTINIQUE

Echelle de Cinq Lieues

# Sailing Wolfe to Quebec ST LAWRENCE RIVER, NEW FRANCE, 1759

This chart is one of a sequence in the pilot book of John Veysey, a navigator on board HMS *Neptune*, which illustrate successive stages of a voyage up the St Lawrence River. The Neptune was the flagship of Admiral Saunders, and was also carrying General Wolfe.

These two commanders had been instructed by the Prime Minister, William Pitt, to work together in an ambitious amphibious campaign. Saunder's Navy was to convey Wolfe's troops to Quebec, and then provide ship-based fire to assist the Army's assault on the French stronghold of Quebec City. Its capture in September 1759 would subsequently lead to the fall of Montreal, and to the end of French control of this part of North America.

Veysey had the task of guiding the *Neptune* through dangerous tides and currents. This page of his log shows the last stretch of the St Lawrence River, from the Kamouraska Islands at right to Quebec City, at far left. As the river narrowed and grew shallower they encountered obstacles, especially in the Traverse, an area of islets and shoals between the island on the right, Île-aux-Coudres, and the larger Île d'Orleans at left. The fleet anchored here in late summer, and the British laid siege to the city, until the night when Wolfe's men mounted a surprise attack. By climbing a tall cliff, British troops gained the Plains of Abraham where they successfully fought the French the next day.

Wolfe's military coup captured the public imagination, although for him this was a one-way voyage, as he died in the battle. Admiral Saunders came home to be heaped with honours, and he recommended Veysey for promotion, as he had been 'very active and serviceable'. We know that Veysey gained his lieutenancy the following year, and much more about his naval service, because several boxes of his personal papers were filed as exhibits in a case

in the Court of Chancery. They contain this pilot book, a number of his manuscript ships' logs, and appointments to command vessels, which together supply an overview of his naval career for a period when official records of service are scant.

That these naval papers were kept is remarkable, as the court case appears to have been about the disposal of Veysey's Devon property after his death in 1808, aged 74. Other papers suggest that he had become a gentleman farmer in the last decades of his life, acting as Justice of the Peace and churchwarden. Among mortgages and fire insurance policies lies a newspaper clipping with notice of a horse auction. An account of the sale of Veysey's household effects lists his books, a Wilton carpet and a mahogany dining set in his house at Brampford Speke. The insight into Veysey's career, and this light thrown on his later years, are made possible by the chance survival of these records in a court of law.

FACE IN THE FIGURES: John Veysey wrote by hand manuals on geometry and on 'Arithmetick Vulgar and Decimal', to which he added illustrations – perhaps even a self-portrait?

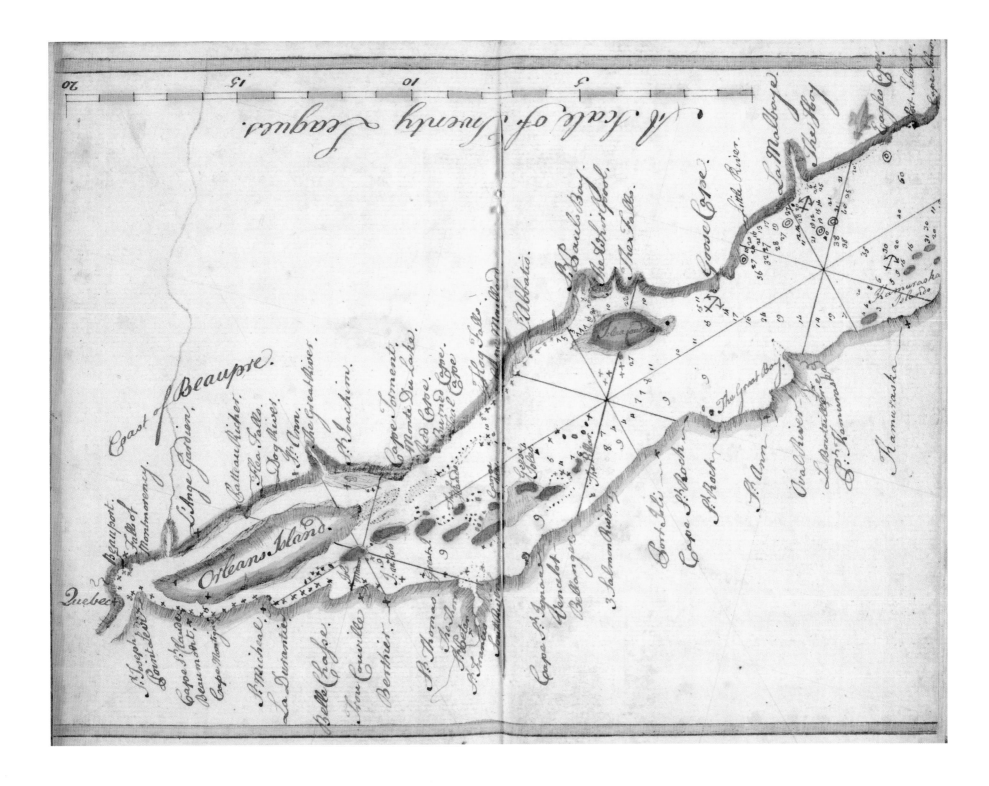

20    15    10    5

Coast of Beaupre.

Quebec

Beauport.
Fall of Montmorency.
Large Garden.
Catteau Richer.
Plea Falls.
Dog River.
St. Ann.
The Great River.
St. Joachim.

Cape Torment.
Monte Du Lake.
Red Cape.
Burnt Cape.
Lost Cape.
Hog Falls.

Orleans Island.

St. Joseph Pointe leve.
Cape St. Laurt Beaumont.
Cape Montg.
St. Michael.
La Durantie.
Belle Chase.
Trou Cuille.
Berthier.
St. Thomas.
The River Bushka.
Cape St. Ignace Vincelot.
Bellange.
3. Salmon River.
Port Jolt
Cape St. Roch.
St. Roch.
St. Ann.
Riva River.
L. Bouteillerie.
St. Kamuraska.

St. Abbatis.
St. Pauls Bay.
The Whirlpool.
The Falls.
Goose Cape.
Little River.
La Malboye.
The Boy.
Eagles Cape.
Cape Salmon.

The Great Bay.
Kamuraska.
Kamuraska Island.

# Discoveries of the *Resolution*

The Age of Enlightenment brought about a change in European thought and art. Mapmakers were no longer interested solely in trade and money, but sought to expand human knowledge about the world. Exploration in new areas of the globe led chart makers to devise new ways to show their discoveries, in areas that had been simply blank spaces or unrelated decoration on earlier charts and maps. The circular chart opposite is a fine example of this phenomenon. It forms the frontispiece to Captain James Cook's log of HMS *Resolution* on her second voyage, which lasted from July 1772 to July 1775. Cook was given secret instructions from the Admiralty to make 'farther discoveries towards the South Pole' (shown at the centre of this chart) and his findings during this voyage would indeed prove that the fabled great southern continent, posited for centuries as a counterbalance to the North Pole, did not exist.

The chart employs an unusual projection to represent the southernmost part of the round Earth clearly on a flat sheet of paper. It gives us the impression of looking at the base of an upturned globe. Africa sits at the lower edge and New Zealand at the top. The oceans range around these sketchy landmasses, through which the ship's route is tracked, with dates to differentiate passages in the same seas. A more conventional portion of chart is placed at the top of the page, offering a continuation northward of part of the chart below. The decorative presentation shows the influence of Classical culture. Supporting the world are two allegorical figures drawn by William Hodges, the expedition's artist. These depict Labour (older and tired-looking) and Science (an energetic youth wearing dividers as a headpiece).

Captain Cook's voyages of exploration are recorded in the archives among Admiralty correspondence, and within ships' logs kept by himself and by his officers. These documents contain many charts, including detailed surveys of the specific islands and parts of the coast where the ships ventured. They provide clear evidence of how Cook laboured steadily in an endeavour to improve knowledge about the world's oceans. He advanced the cause of scientific chartmaking, with a careful use of running surveys to fix coastlines accurately. His ship was equipped with one of John Harrison's chronometers on trial, which he praised in his log: 'We have received very great assistance from this usefull and valuable time piece'. Thus the voyage played an active part towards a solution for the problem of calculating longitude. The final lines of Cook's last letter from his last voyage in 1778 – received in London a year after his death – read 'whatever time we do remain shall be spent in the improvement of Geography and Navigation'.

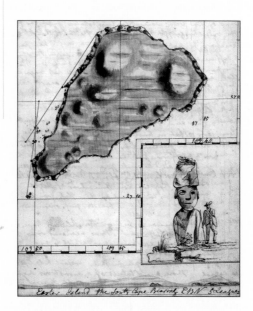

EASTER ISLAND: This watercolour plan and view come from a page of the log kept by Joseph Gilbert, master of HMS *Resolution*. Gilbert also drew one of the island's statues with a naval officer standing beside it, to convey a sense of its monumental size.

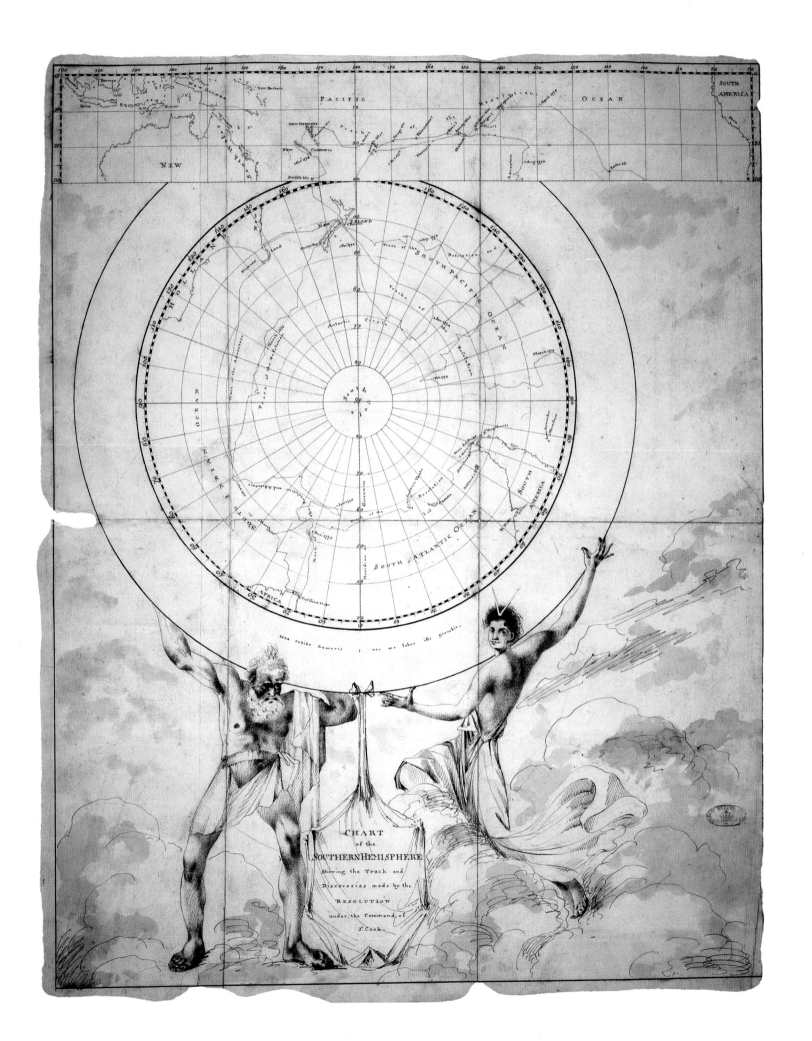

CHART
of the
SOUTHERN HEMISPHERE
Shewing the Track and
Discoveries made by the
RESOLUTION
under the Command of
J. Cook.

# After the Mutiny

The first time that Captain William Bligh encountered these islands – and named this 'discovery' after himself – he was not in circumstances that would allow him to stop to chart them. He had been cast adrift in an open boat, thousands of miles from the nearest settlement, and had to navigate to safety with 18 of his crew who remained loyal to him after the notorious mutiny on HMS *Bounty*. The broken line on this chart, which runs diagonally from lower right to top left, tracks this journey in the *Bounty's* launch over four days from 4 May 1789 – just a week after their ignominious expulsion from the commandeered ship. The larger islands at top left, with green-tinted contours, were also observed at this time. The strait between them is still named Bligh Water, although the archipelago is now known in English as Fiji.

Bligh returned to these islands in 1792, several years after the Bounty fiasco, with plans to 'verify my observations during my distressing Voyage', from the pleasanter quarters of his captain's cabin on HMS *Providence*. The solid line on the chart zigzags about the islands, showing where the ship followed a glimpse of land when the cloud and rain lifted. Bligh – whose signature appears under the chart title – only recorded what he saw for himself or what was reported to him as seen from the top of the mast head. This gives the chart an unfinished effect. Despite his desire to complete a full survey, this was never to be. 'If I had had a Month to spare I would have completed it myself', he remarks in his ship's log.

It was not just the weather that was against Bligh; time, too, was not in his favour. His main task was to transport breadfruit plants from Tahiti, where they grew plentifully, to feed the burgeoning colonies in the West Indies. His log shows that several thousand plants were put on board in July 1792. Bligh was evidently still visiting the plants daily by mid August, when this chart was drawn, as he notes that they were 'in charming order'. To keep them like that, he would need to deliver them to their destination as soon as possible, through a difficult route and 'a contrary Monsoon'. Although the survey remained incomplete, the botanical mission proved thoroughly successful, as he lost only a few hundred plants.

This episode was part of a long naval career for Bligh. He had learnt much about making good surveys from Captain Cook on board the *Resolution* in his youth, which he in turn passed on to Matthew Flinders, a young midshipman on this voyage of the *Providence* who would later become the first man to circumnavigate Australia. Bligh served with credit under Nelson at the Battle of Copenhagen in 1801, for which he drew a plan held by the archives. Whatever history's verdict might be on Bligh as a naval officer, he was a superb navigator and chart maker.

ISLAND VIEW: Cook had also sailed near Fiji. This view of Turtle Island was drawn by the *Resolution's* master Joseph Gilbert.

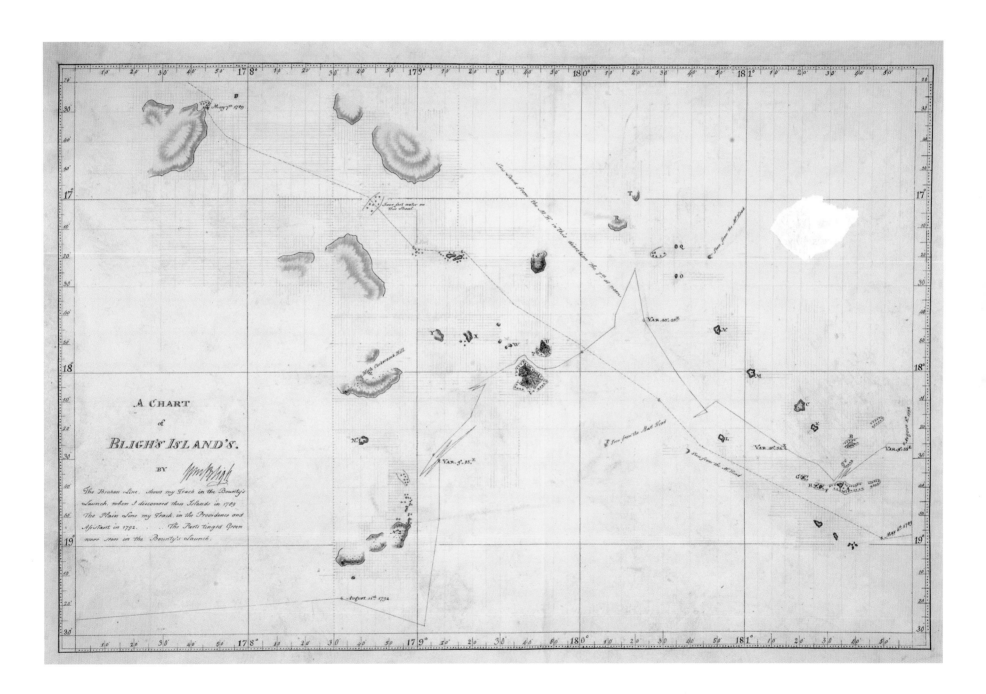

A CHART

of

BLIGH'S ISLAND'S.

BY Wm Bligh

The Broken Line shews my Track in the Bounty's
Launch, when I discovered these Islands in 1789
The Plain Line my Track in the Providence and
Assistant in 1792. The Parts tinged Green
were seen in the Bounty's Launch.

# Battles of the Nile

There are places in the world that are significant at very different points in history. Alexandria is one of them. Founded by Alexander the Great, it became one of the foremost cities of the Classical world, with its famed library as a treasure-house of knowledge. Centuries later, this area was the scene of major land and sea battles during the French Revolutionary Wars. The sites of many of these are recorded on this overview chart of British operations between 1798 and 1801. It was made, however, in the initial stages of a diplomatic crisis 40 years later, which would result in action by the Royal Navy off these shores once again. Despite the immediate purpose of the chart, it was drawn in a way that makes evident all these layers of history.

Only the right-hand side of a much longer chart is illustrated here, with Alexandria just off the left-hand edge. A French fleet landed to the west of the city in May 1798 to invade Egypt, as the first step in Napoleon's plan to force the British out of the war with France, by a campaign against India. At the far right is shown the Battle of the Nile. A line of ships represents the French fleet in formation, as encountered by Nelson's fleet on 1 August 1798 ('1898' on the chart is an error). Dotted lines show how the British ships trapped the French in a pincer movement. The French flagship *Orient* in the centre of line was 'blown up' – this was the famous 'burning deck' on which the young son of Captain Casabianca stood firm, until flames caused the ship's ammunition to explode.

The dates on the chart then move on to March 1801, and the scene in the middle of the top sheet, where a note reads 'British Landing … led on by [Sir Ralph] Abercromby'. A task force of about 17,000 British soldiers were ordered to dislodge the French army, which had remained in Egypt under Napoleon after their fleet's destruction. Fierce action took place in this area of sand dunes and palm trees. At the far left of the lower sheet, crossed swords next to the ruins of Caesar's Camp mark the 'Defeat of the French'.

The chart is dated 30 January 1840. It was sent to the Foreign Office that April in a secret letter from the British Ambassador to Vienna. There, discussions were being held about the 'Oriental Crisis', caused by the Ottoman viceroy in Egypt, who sought to rule in his own right. Britain, in alliance with the Austrian Empire and other European powers, decided to support the Ottoman Empire against him. The far left of the chart (not shown here) bears notes on Alexandria's defences, detailed in an accompanying report recommending how best to attack them. These papers were forwarded to the Admiralty to assist in planning the new campaign – informed by the lessons of the past.

CLEOPATRA'S NEEDLES: The chart notes 'ruins passable for all arms'; a practical military attitude to the remnants of ancient civilisations, such as these ruins of Ptolemy's Palace, and 'Cleopatra's Needles'. One of these obelisks arrived in London in 1878; the other in New York in 1881.

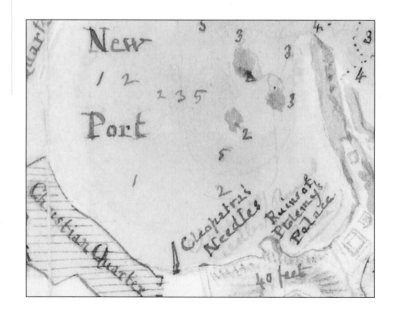

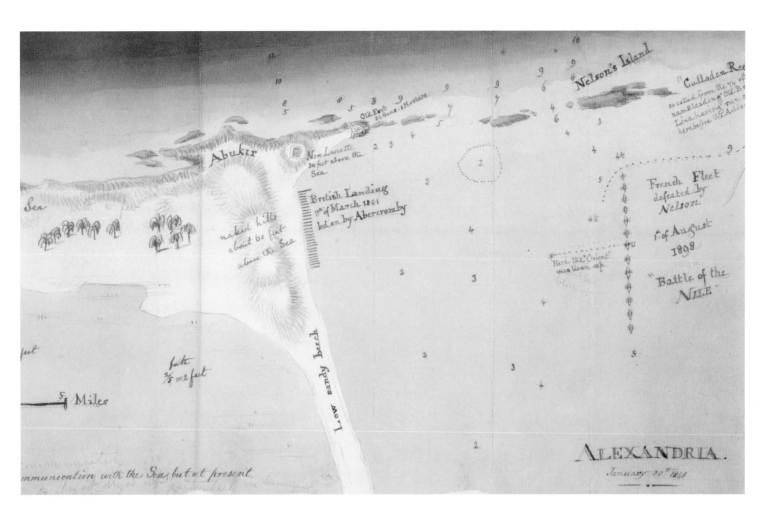

Sea

Abukir

naked hills
about 60 feet
above the Sea

Old Fort
24 Guns 1 Mortar

New Lunette
30 feet above the
Sea.

British Landing
8ᵗʰ of March 1801
led on by Abercromby

Nelson's Island

"Culloden Reef"
so called from the 74 of
name leading the Br
Line having run
heretofore it

French Fleet
defeated by
Nelson

1ˢᵗ of August
1898

"Battle of the
NILE"

Here the "Orient"
was blown up

Low sandy beech

scale
⅜ = 2 feet

Miles

communication with the Sea; but at present
...

ALEXANDRIA.

January 30ᵗʰ 1840

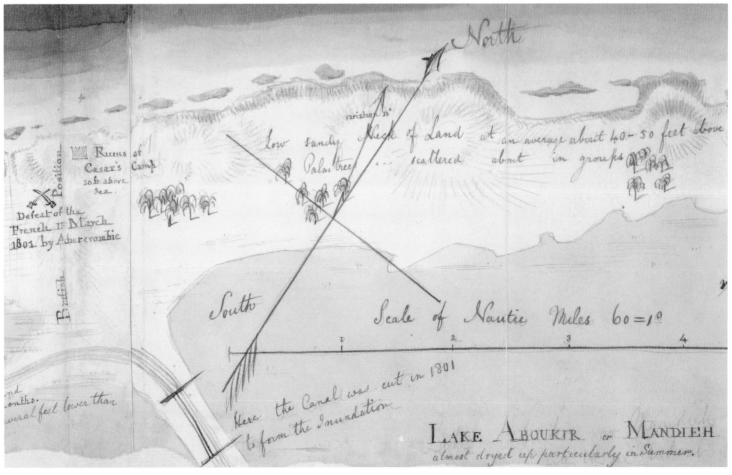

North

Ruins of
Cæsar's Camp
20 ft above
Sea.

Position

Defeat of the
French 15 March
1801 by Abercrombie

British

variation 15°

Low sandy Neck of Land     at an average about 40 - 50 feet above
Palm trees ... scattered     about in groups

South

Scale of Nautic Miles 60 = 1°

1     2     3     4

Here the Canal was cut in 1801
to form the Inundation

LAKE ABOUKIR or MANDIEH
almost dryed up particularly in Summer.

and
months.
...several feet lower than

# White on white

In the far north, the sea freezes, and when snow and frost cover the land, all turns to white. Much of this large hand-drawn chart is also white, where areas of paper are blank. Here, we have selected some details which plot the course of the explorer and naval officer William Edward Parry in the Arctic. His was one of the first 19th-century British voyages to the far north of what is now Canada. The blank spaces symbolise the unknown regions which surrounded Parry and his men, as they sought to fix bearings and chart coastlines that appeared occasionally among the shifting ice-floes when the fog lifted.

After the Napoleonic Wars, the Royal Navy no longer needed so many men on active service. John Barrow, Secretary to the Admiralty – commemorated by Barrow Strait on this chart – sent officers to search for the North-West Passage, which it was hoped would give European trade a short cut to the Far East. The young Lieutenant Parry served with Captain John Ross on an Arctic expedition in 1818. Parry then commanded two ships, HMS *Hecla* and HMS *Griper*, to explore beyond the point reached by Ross, and this chart records discoveries on their voyage from 1819 to 1820. The ships sailed up Baffin Bay, then westward through Lancaster Sound – which Ross had believed to be blocked by mountains – to Melville Island. They overwintered there, intending to go further the next spring, but were thwarted by pack ice.

The chart is full of detail of this area, hitherto wholly unexplored by Europeans. Near Liddon's Gulf – named for the Griper's captain – plentiful game fed on abundant grass, sorrel and moss saxifrage. At the entrance to Prince Regent Inlet were seen many large black whales, narwhals and seals. Parry hoped this would be an easier way to the west, but it proved to be a dead end. A bottle buried on its shores is marked by a cross on the chart.

The harsh climate and other hazards hampered their attempts to put lines on the chart. Long Island's position and outline were 'very uncertain owing to ... indifferent observations ... in consequence of thick weather'. A faint pencilled note by Parry explains 'This coast line not inserted, in hopes of obtaining more accurate angles'. The shoals in North Cove were dangerous, while in Hecla Bay they encountered 14 feet of ice, in an area of snowy plains.

The perils of this kind of voyage were extreme: a ship could be crushed by ice or holed on an iceberg, or explorers might run out of supplies and die of starvation, as John Franklin's ill-fated final expedition of 1845 would grimly demonstrate. Although Parry did not find a complete way through the North-West Passage to the Pacific, this was still perhaps the most successful voyage in the history of its exploration, and the channel was named after him. This major discovery is recorded on this chart among the dotted lines and blank spaces.

CONVEYANCE FOR PEMMICAN Searches for the lost expedition of Parry's friend Franklin continued for many years after 1845. In 1850 Parry suggested that searchers use a sledge made from four Lapland snow-shoes, a design that he had successfully deployed in the Arctic in 1826.

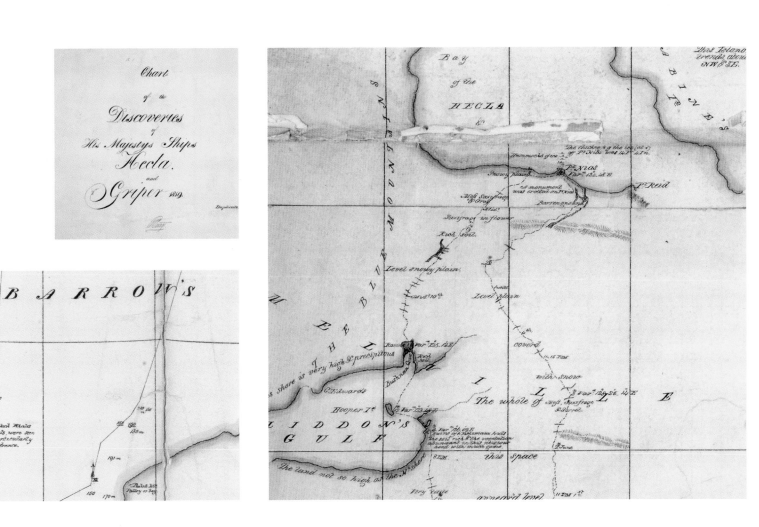

Chart
of the
Discoveries
of
His Majestys Ships
Hecla.
and
Griper 1819.

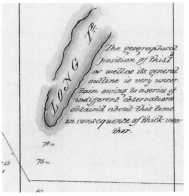

The geographical position of this, as well as its general outline is very uncertain owing to a series of indifferent observations obtained about this time in consequence of thick weather.

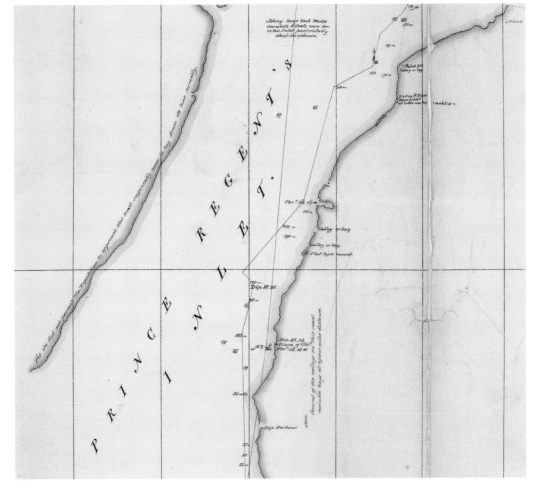

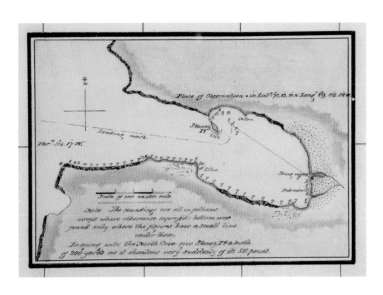

Note. The soundings are all in fathoms except where otherwise expressed; bottom was found only where the figures have a small line under them.

# Gates to St Petersburg

KRONSTADT, RUSSIA, 1851

In the autumn of 1851, HM Steam Frigate *Odin* conveyed Sir Hamilton Seymour from London to take up his appointment as the British ambassador to the Russian imperial court in St Petersburg. The diplomatic purpose of this voyage meant that the commander of the *Odin*, Sir William Wiseman, was treated as an honoured guest in Russian waters. When he anchored in Kronstadt, a lieutenant in the Russian navy was sent by the governor to attend upon him. Wiseman took the opportunity to be shown round this strategically-sited island, at the eastern end of the Gulf of Finland. The seat of the Russian Admiralty and base for their Baltic fleet, it guards the way to St Petersburg, about thirty kilometres to the east.

Wiseman found that 'Notwithstanding this apparent civility … it was pre-arranged what we were to be shown'. The British party saw only a few of the latest Russian ships, which Wiseman guessed was because their main fleet was outdated compared with the steamship *Odin*. They were, however, given a tour of the island's fortifications, of which the Russians were especially proud. Wiseman made a detailed report on these, together with what he could find out about the workings of the Russian navy. This was only able to sail in the four or five months when the northern seas were not ice-bound; the sailors effectively became soldiers for much of the year, which Wiseman thought might affect their seamanship.

This chart of Kronstadt with encircling panorama depicts what could be seen from the *Odin*'s anchorage, denoted by an anchor sign above the 'b' in the legend 'Channel to St. Petersburgh'. This lies in the chart's centre, heading towards the top of the page, with 'Cronstadt' at top left. The chart shows soundings, sandbanks, the location of the forts around the channel, and the number of their guns. These defences are also shown in view at the edge, as they appeared from on board ship. Other buildings in the town with distinctive silhouettes are numbered, and named in the key. These include the English and Catholic churches, the cathedral, the British consulate, the Custom House and the Admiralty building. The chart and four fort views were made from memory, after the visit, by Augustus Whichelo, the *Odin*'s Second Master – a remarkable feat given that his eyesight was failing at the time.

Wiseman sent his report with Whichelo's chart and illustrations to the Admiralty in January 1854, stating in a covering letter that 'from the present aspect of affairs, they may prove useful'. This alludes to the fact that Russia and the Ottoman Empire went to war in October 1853, and Britain seemed likely to enter the conflict, against Russia – which she did, in March 1854. These papers were apparently sent to Sir Charles Napier, who commanded the British fleet in the Baltic. The fact that Napier avoided making an attack on Kronstadt may have been due to Wiseman's intelligence that the forts shown on this chart were well-nigh impregnable.

FORT PRINCE MENZIKOFF: Whichelo's watercolour view conveys the strength of the granite fort, with gun-ports set at a height to rake enemy ships in the channel.

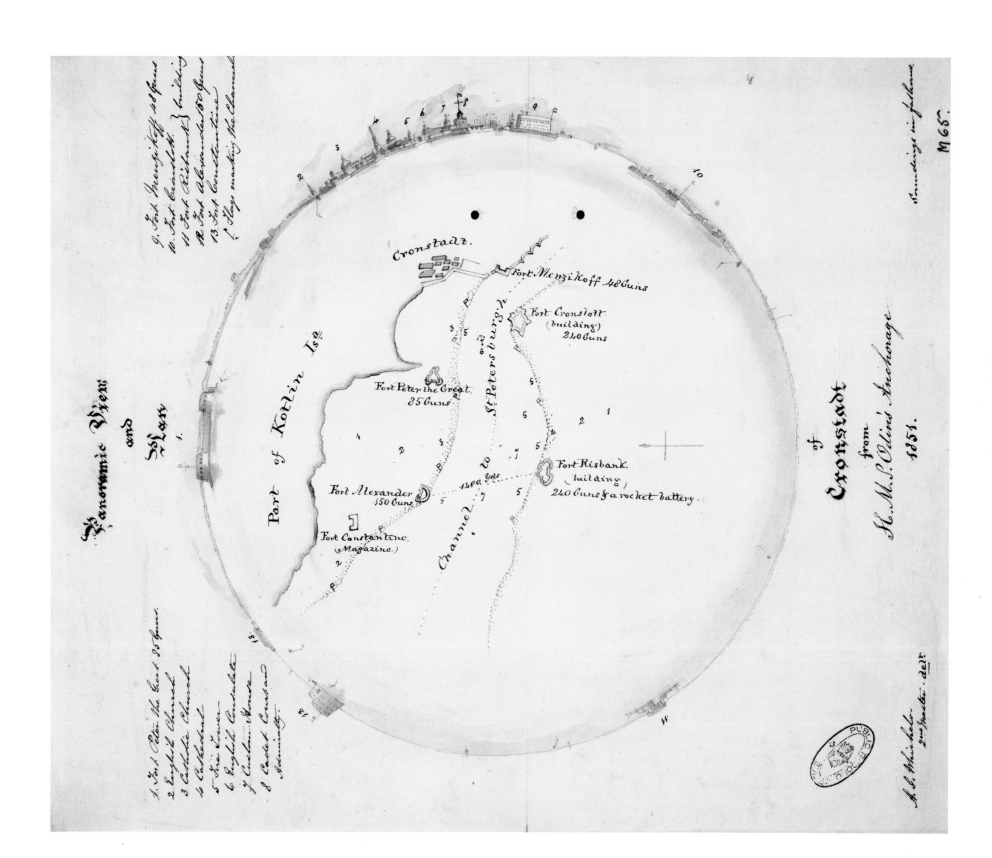

Panoramic View and Plan

Part of Kotlin Id.

Cronstadt.

Fort Menzikoff 48 Guns

Fort Cronslott (building) 210 Guns

Fort Peter the Great 85 Guns

St Petersburgh

Fort Risbank (building) 240 Guns & a rocket battery

Fort Alexander 150 Guns

Fort Constantine (Magazine.)

Channel to 1200 Yds.

of

Cronstadt

from

H.M.S. Odin's Anchorage

1851.

Soundings in fathoms

M 65.

# CHAPTER SIX

# NEW WORLDS:
# exploration
# & the colonies

The group of islands that make up the British Isles long ago turned its attention beyond its shores, and so the archives contain many maps of places overseas. Some of these appear in other chapters where they portray foreign towns, landscapes and seascapes, or the specific focus of military action. This chapter offers a whirlwind tour of maps that illustrate the story of the rise and fall of empire. The first map relates the tale of the first English colony in the New World in 1585. The last records the end of British rule in India after the Second World War. Maps accompanied the different phases of colonial enterprise such as exploration, colonisation, boundary definition, and the practicalities of power on a global scale.

In the period of prelude to empire, Europeans explored beyond the boundaries of the world known to them, and they encountered peoples already living in far-off places. Maps in this chapter portray how a native American, a Maori priest and an Aboriginal Australian saw their worlds (pages 175, 183 and 185). Despite the fact that America

was a continent inhabited by numerous indigenous peoples, early European visitors perceived it as a New World available for discovery, occupation and exploitation. The map of Raleigh's colony on page 169 is roughly drawn, but soon more detailed surveys were made to facilitate settlement, as we see on a map which records a journey in the region of the native Iroquois in what is now upstate New York (page 173). Other European powers were also seeking to gain control of the same areas, so a survey might take the form of military reconnaissance, which could be accompanied by danger, as we find in the story behind the map of Creek Indian territory on page 179.

From the 17th century the map of empire began to be unrolled. Trade was a major reason for overseas expansion, with the aim of sourcing raw materials and capturing monopolies in new markets abroad. A map made for the Royal African Company showing commercial routes along the West African coast is on page 171. Trading stations of the European powers proliferated there, as first

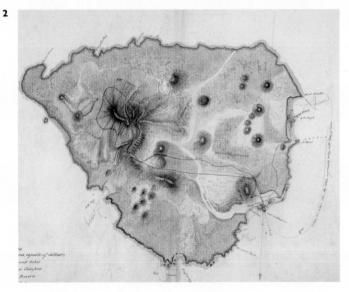

**1** EARLY NEW YORK: The strategic location of the then small town of New York, on the southern tip of Manhattan Island, is evident on this chart-map of 1700.

**2** ASCENSION ISLAND: This isolated volcanic island between Africa and South America was a British naval base called a 'stone frigate'. Despite an unpromising appearance of cinder, ashes and lava, this 1829 plan notes water pipes, roads, and land capable of cultivation.

**3** A PLANNED CITY: This portion of an 1863 plan of Victoria, capital of British Columbia, shows a townscape planned around the Hudson's Bay Company wharf, with an indigenous village opposite, and streets laid out on a grid pattern.pipes, roads, and land capable of cultivation.

gold and then slaves were exported; a slave fort is shown on page 177. Investigation of Africa's interior came later, in the 19th century. David Livingstone recorded that no white man had been seen before in the areas of East Africa he traversed on an 1859 expedition, mapped on page 187.

Europeans took to the new colonies notions of statehood which entailed formally defining boundaries, especially where territories of different powers adjoined. Opposite are examples of maps showing borders between states in New England, and between provinces in South Africa. Maps offered a means of defining frontiers that could be combined with written documents to locate the border lines precisely. Once state boundaries were in place, it was a question of dividing the area according to European notions of landholding. Many maps record lands granted; an example on page 181 shows the problems caused where more than one power issued grants in the same area.

Some maps were strictly practical – for instance, those that recorded details of navigability of rivers or soil fertility, to facilitate further exploration or settlement. Others are designed to convey a marketing message. The decoration on an early map of Virginia (see the detail on page 169) was designed to promote the colony and encourage settlement, with a royal coat of arms connoting power and prestige. Once established, colonies needed to be supported, and a poster from the heyday of the British Empire was intended to encourage the British to buy goods from her colonies and dominions (page 189).

From the 1930s onwards, and especially after the Second World War, there was movement towards decolonisation. A map on page 191 indicates the potential complexity of this phase, especially where Britain had ruled an area for a long time. Thus the map of empire was gradually rolled up. Yet maps of the colonial era remain in the archives, continuing to inform about major matters such as international boundaries, and to fascinate with vistas of worlds that may be long gone but appear fresh with interest to us today.

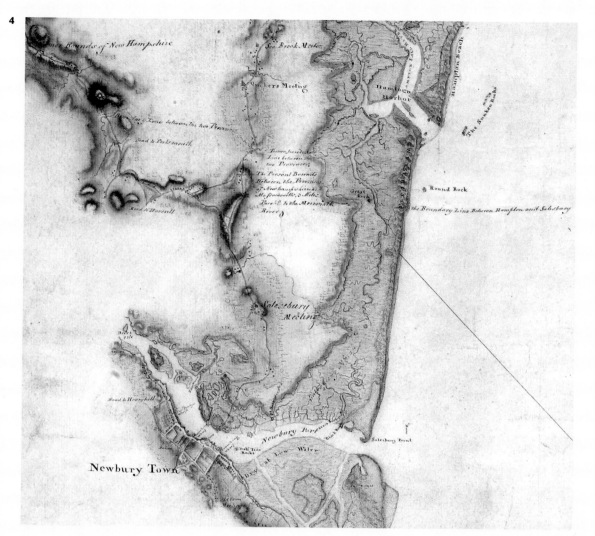

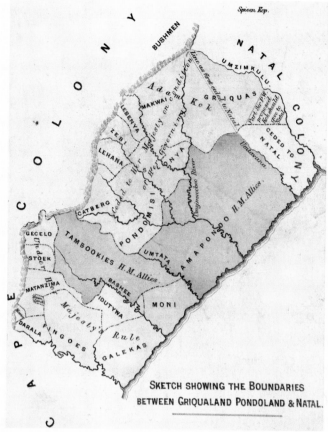

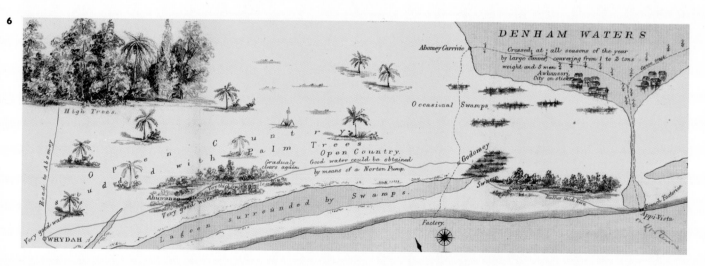

**4** NEW ENGLAND BOUNDARY: This map was drawn in 1760 by Charles Blaskowitz, Deputy Surveyor General, and shows the boundary between Massachusetts and New Hampshire.

**5** SOUTH AFRICAN BOUNDARIES: Complex boundaries in central South Africa are shown on this map of 1885, which differentiates between lands 'Under Her Majesty's Rule', those of 'H.M. Allies', and lands ceded to various parties.

**6** AFRICAN LANDSCAPE: This map shows the route of an expedition in 1876 in Benin, on the West African coast. Noted are a lagoon surrounded by swamps, tree cover, and a 'city on sticks'.

# Land of great red grapes

In the late 16th century England set her sights on America. Descriptions sent home by early expeditions verged on the lyrical, which encouraged English settlement. An account sent to Sir Walter Raleigh in 1584 spoke of the land shown here as 'so full of grapes', where 'in all the world the like abundance is not to be found'. Raleigh held letters patent from Elizabeth I to claim lands in the New World, and this prompted him to send a party of settlers there in the following year. 'Great store of great red grapis veri pleasant' is noted on the far right of this sketch of the Roanoke settlement, which is believed to be the earliest English map of North America made from direct observation.

The map is drawn as though approaching the coast from the Atlantic, with the North Carolina Outer Banks along the lower edge. A note by the gap between the first two islands on the left shows 'where we arivid first'; here the party sailed to shelter inside the Banks. Above that is Pamlico Sound, with Albemarle Sound to the right in which lies King's (now Roanoke) Island, where the colony was set up. Other islands and the names of native villages are marked, along with places where useful plants grew: oak galls for making ink, 'the roots that dieth red' (madder), and 'the grase that berithe the silke' (a forage crop). This was all valuable information for the prospective coloniser.

This map was probably sent to London with a letter dated 8 September 1585 from the new colony's governor, Ralph Lane. The settlers had arrived in late summer, which was grape harvest time – also the season when the glowing report of the previous year was written. The enticement of plentiful delicious food was not available all year round, and the colonists would find the living much less easy in winter. They had lost most of their provisions on the journey, and had arrived too late in the year to plant crops. They lacked nets to take advantage of the 'great store of fishe' shown below the grapes on the map, so they had to live off the land – and off the generosity of the native people already dwelling there.

Lane's letter was addressed to Sir Francis Walsingham, Queen Elizabeth's spymaster. Manuscript maps such as this one were a way for the English queen to control knowledge of new overseas lands, since they had limited circulation among her close advisers. In parallel Spain, which acquired a large gold-bearing empire in South America, kept its master map of discoveries, the padrón *Real*, in great secrecy. 'A description of the land of Virginia' is written on the back of this map; Raleigh had agreed to name the first new settlement thus, in honour of the Virgin Queen. This first attempt failed, but the name lived on nearby, in what became the first successful colony, Virginia.

VIRGINIA ENGRAVED: This 1608 map is based on explorations of John Smith, a founder of Jamestown in 1607. Designed for wide circulation, it conveys a strong message promoting the new colony. The arms of James I give connotations of royal blessing, while the native figure aimed to stimulate interest.

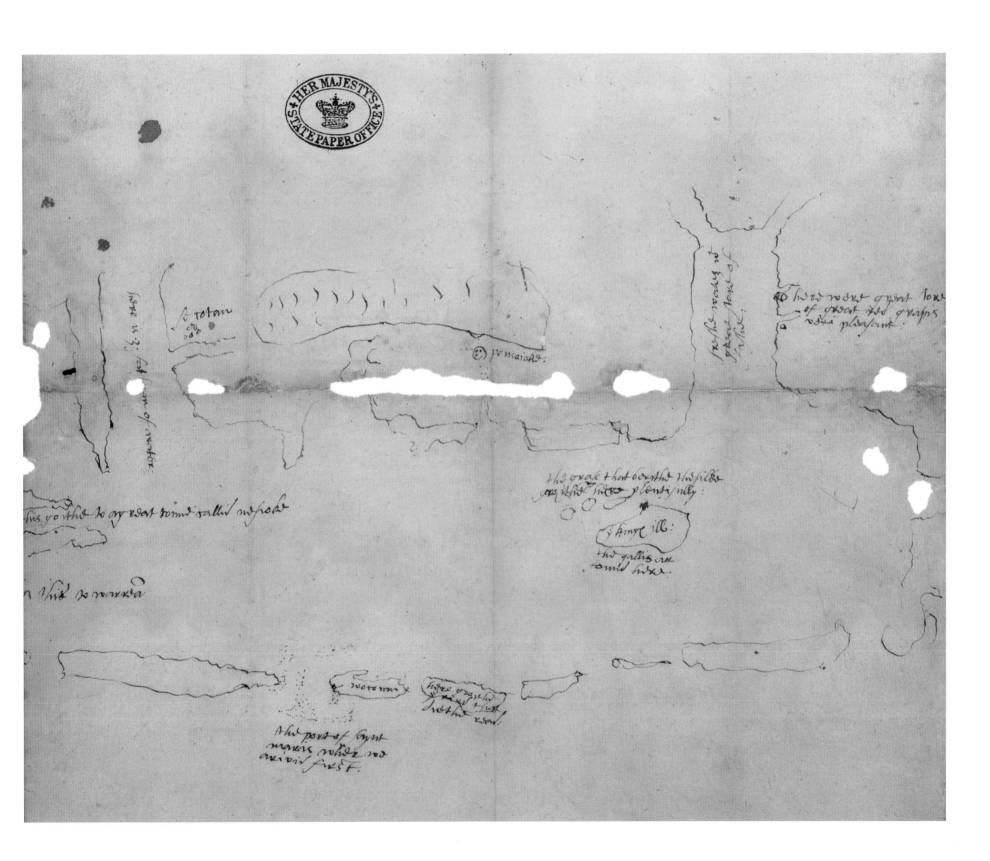

# Elephant and castle  WEST AFRICA, c.1680

The coat of arms at the top of this engraved map is that of the Royal African Company of England, and below it is a dedication in Latin to the Company's governors and principal investors. Our example of the map was used as the frontispiece to a bound book in which were written texts of the Company's official papers, including its 1672 charter. This conferred a trading monopoly, aimed at capturing the lucrative market in gold, ivory and slaves – for each of which stretches of these African coasts were named. The Company transported thousands of slaves to work in the West Indian sugar plantations which grew up in the later 17th century to meet demand in Europe. A sailing route is shown from Cape Verde across the Atlantic, to South America at the lower edge.

The map is entitled 'A chart of the Seacoasts from the Landsend of England to Cape Bona Esperanca'. East is at the top, with the Cape of Good Hope at far right, and southern England at the far left. The meridian of the Lizard is used, rather than the modern zero point at Greenwich; it runs from near Lands End, horizontally under the feet of the seated figures. The map is drawn to make it appear that by sailing south, like the ship pictured, one soon encountered 'Maroco', by the Barbary Coast. Continuing along the African shoreline, a Company ship would find all of its trading places, such as Benin, Gabon and Angola. Guinea provided Company gold to the Royal Mint for the coins given that name, each stamped with an elephant, while the Gambia was a source of ivory (or 'elephants' teeth').

Like the chart on page 145, this is a plate from John Seller's *Atlas Maritimus*. In 1671 he was made Royal Hydrographer by Charles II, whose brother James, Duke of York, was a governor of the Royal African Company. It was said that James intervened when Seller, a non-conformist, had been accused of conspiracy against the King; whatever the truth of this, Seller also dedicated other works to the Duke. Although Seller's royal appointment gave him a privileged position to publish nautical atlases, the title 'hydrographer' did not mean that his charts necessarily focused on navigation. This one, although it does show detail of coastlines, seems to be more a marketing exercise for the Company and for Seller himself.

European mapmakers at this date knew very little about the interior of Africa. As the satirist Jonathan Swift remarked:

*So geographers, in Afric maps,*
*With savage pictures fill their gaps,*
*And o'er unhabitable downs*
*Place elephants for want of towns.*

On this map, the technique of disguising lack of knowledge about the middle of the continent by adding carefully-placed animals is given a twist. The space is filled with decoration, including the Royal African Company's emblem of an elephant saddled with a castle.

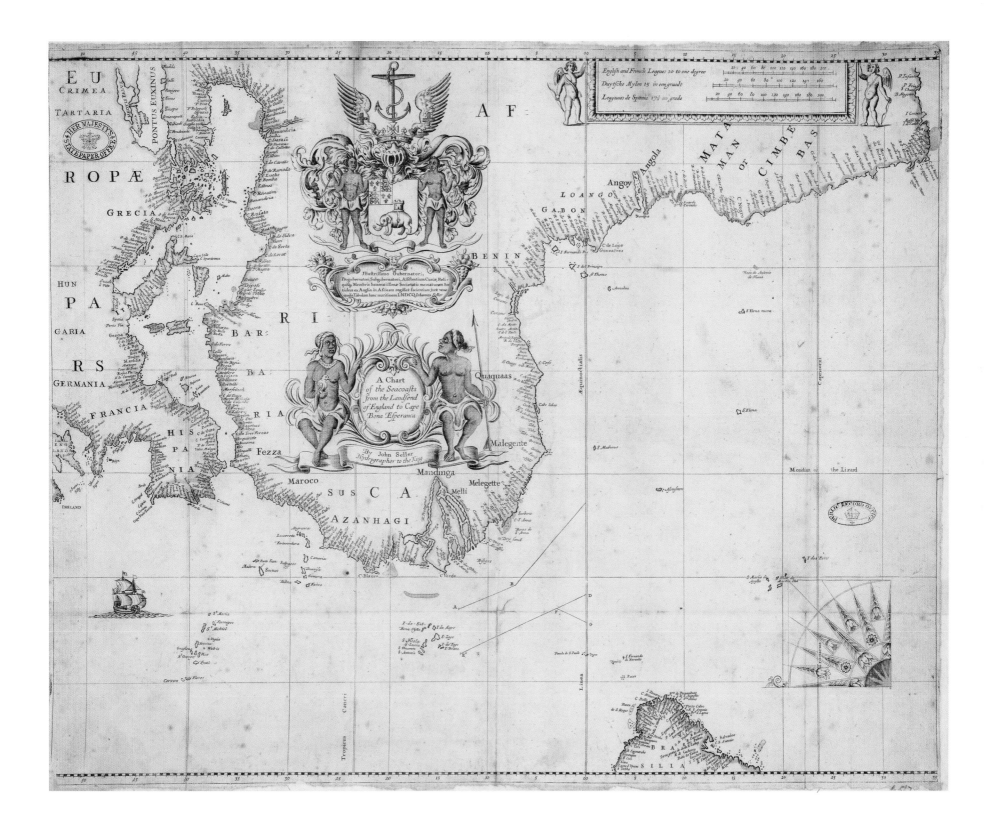

# 'Journey to the 5 Indian Nations'

There is often little visible difference between early maps made for military reasons and those concerned with establishing a colony. This map was made by Colonel William Römer, an engineer in the British army, who was sent to New York in 1698 to oversee the fortifications of the city and province. It shows his journey to the interior of what is now upstate New York, to survey places for defence against the French, who also wanted to control this area. These lands to the north-west of New York were largely unmapped by Europeans. They were territories of the Five Nations, a confederacy of native Iroquois, with whom the British sought an alliance against the French. Thus the map incorporates aspects of exploration, military intelligence, diplomacy, general topography and much information about the tribes.

The area shown lies between New York City, off the lower edge of the map, and the Great Lakes region. At right, the Hudson River flows through and beyond Albany, which lies at its fork with the Mohawk River (here named the Maquas). Heading westward past Lake Oneida we come to 'Cadragqua Lake' (now Lake Ontario). This is joined to Lake Erie by the Niagara River, on which lay a 'great fall' – an early mention of Niagara Falls, then recently discovered by the French. The areas with little detail are those where Römer used others' descriptions. In the empty spaces he drew an Indian village, a turkey, beavers and bears.

In a detailed note below the title, Römer explains that he actually visited only three of the Five Nations. From Albany he journeyed west to the Mohawks, the First Nation, and thence to the 'Onyades' (Oneida) and the Onondaga, the Second and Third Nations. He then relates mysteriously that he was 'stoped and could not procede any forther for sum important reasons'. Whatever these were – perhaps to do with local politics – he did not continue further west to visit the Cayuga and the Seneca, although he includes their lands on his map. Instead, he turned north to reach part-way up the Oswego River in the direction of Lake Ontario before, his provisions low, he turned back for Albany.

Although Römer was unable to visit the Fourth and Fifth Nations, he was still among the first Europeans to visit the Iroquois, and to map the locations and names of their settlements (denoted here by cabin symbols). He conveys the impression that it was relatively easy for Indians from the Great Lakes to bring furs to trade in Albany, via Lake Ontario and Lake Oneida. His map shows a short portage to transfer goods overland from the latter to the Mohawk River (although he had not himself seen this). In the midst of his explorations, Römer also carried out his military commission. He selected a site, shown by a fort symbol where the Oswego River joins the south-west corner of Lake Ontario, and there a British fort was eventually built.

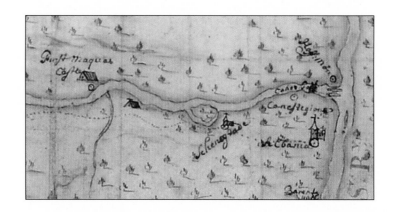

WEST TO THE FIRST NATION: Römer's route overland is shown by 'Read pricked Lines' running below the Mohawk River. From Albany, with its church symbol, he went via Schenectady to three 'Castles' or settlements of the Mohawks. To reach the one shown here, he had to take to a canoe.

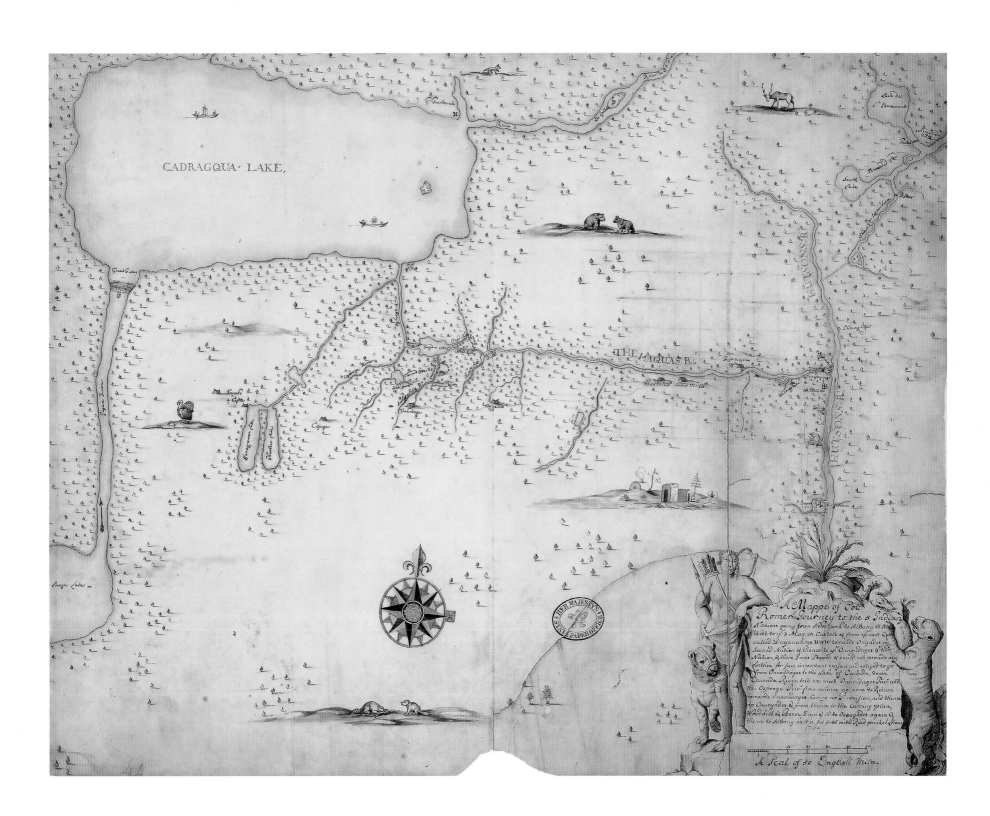

CADRAGQUA· LAKE,

THE MAQUAS R:

A Mappe of Coll
Romers Journey to the 5 Indian
Nations going from New: Yorck to Albany th: Fri
Wde to y: 5 Maquas Casteels & from y: last Cast:
called Daganahage W: SW towards Onyades or
Second Nation & thence to y: Onnoidages & that
Nation, & there Jndos Hoped y: could not procede any
forther for sum important reasons wd: obliged to go
from Onnoidages to the Lake of Canada, down
Canenda River till we meet Onnoidages Rivr add
the Oswego Rivr from whence we were to Return
towards Onnenoages having no Provision and thence
to Onneyades & from thence to the Carring place,
Wod·lill & Oswego Dam & c to Onneydes again &
thence to Albany as it is set forth with Read prickd lines
1700

A Seal of 50 English: Miles

# 'An Indian a Hunting'   CAROLINA, c.1721

The map behind this map was a 'Draught Drawn & Painted upon a Deer Skin by an Indian Cacique', as the title of this map tells us. Although this copy of the chief's map is made on paper, it has kept the shape of the deerskin, with the animal's neck to the right. This is one of very few surviving maps from this period which record the topographical distribution of native Americans in this area. The different peoples are named within circles of varying size joined by lines, which perhaps indicate roads but may represent the state of relations between the various nations. Some have multiple links while others have only one. While circles indicate the Indian areas, straight lines and rectangles are used for places associated with European colonists; Virginia at lower right, and an early town plan of Charleston at left on the eastern seaboard. Was this an attempt to distinguish between the natural world of the Indians and the man-made landscape of the newcomers, perhaps, or simply that these were places of different sorts of people?

This has been widely labelled as 'the Catawba map' because the more numerous smaller circles seem to represent Catawba areas, although the Cherokees and the Chickasaws get a relatively large circle each. Pictographs or stylised drawings show a man hunting a deer, and a larger upright figure with arms akimbo, perhaps

dancing? These and the drawing of the ship in harbour at Charleston all face different directions; it is the copyist's legends added to the map which require the viewer to look at it with south at the top. Here, the east coast is shown on the left and not, as we would expect, on the right. The original deerskin map is not known to survive, so we cannot now tell how the words inside the circles were conveyed on the original, or how much the copyist has added.

The original map was presented to Francis Nicholson as Governor of Carolina, which dates it between the years 1721 and 1725, when he held that post. The map may have been made to show the new English colonial administrator the strategically important links between Indian groups and with the British colonies, or simply as a gift to help cement trading links.

Nicholson considered the map important enough that he arranged to have it copied. One version made on Nicholson's behalf and dedicated to the Prince of Wales is now in the British Library. That map differs slightly in the detail of lines drawn between circles from our version, which he sent to the Secretary of State for the Colonies. Through this map we glimpse the map behind it, and an older tradition of spatial representation, where the map was as much a record of social knowledge as a geographical construct.

HUNTSMAN: Pictograph of a man with bow and arrow leading a horse, from a second map also presented to Nicholson, with a similar title and circular representation of nations, but showing a much wider area of south-eastern North America.

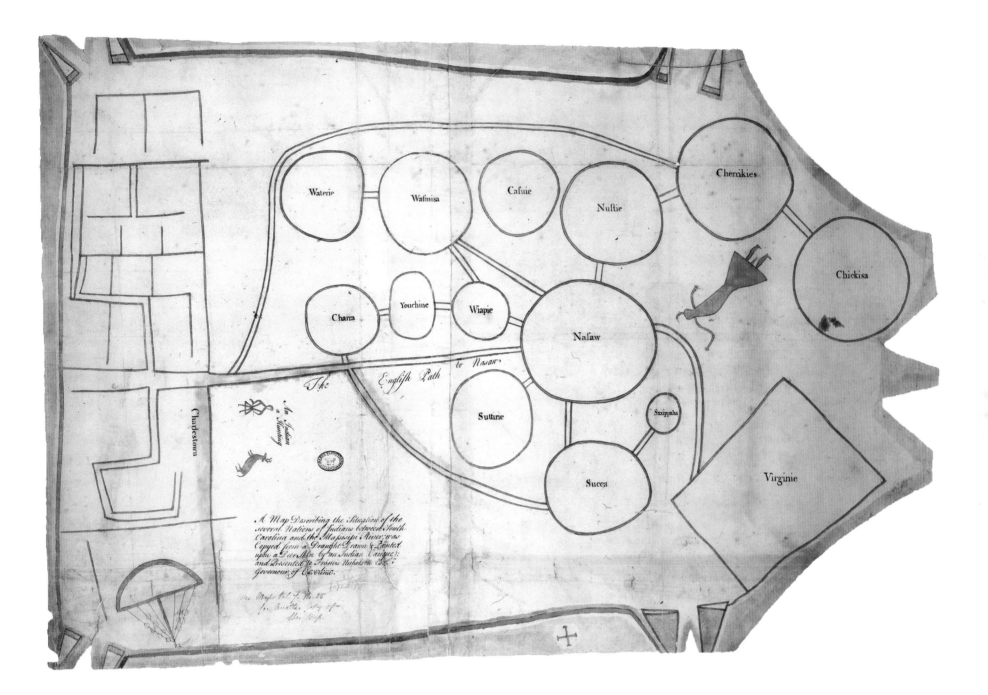

Charlestown

Waterie

Wafmisa

Casuie

Nustie

Cherrikies

Chickisa

Charra

Youchine

Wiapie

Nasaw

Saxippaha

The English Path to Nasaw

An Indian a hunting

Suttirie

Succa

Virginie

A Map Describing the Situation of the
several Nations of Indians between South
Carolina and the Mississipi River; was
Copyed from a Draught Drawn & Painted
upon a Deer Skin by an Indian Cazique;
and Presented to Francis Nicholson Esqr.
Governour of Carolina.

# A slave fort

JAMES ISLAND, RIVER GAMBIA, 1755

In the autumn of 1755, Justly Watson, Director of Engineers, arrived in West Africa. He was there to report on the state of the British fortifications between the Gambia and the Gold Coast. This was a considerable distance, and the tour occupied him until the following May. His maps of eleven forts made on these visits offer a view of how they were set within the surrounding landscape. Each fortress was a centre for trade with its local area and a large hinterland, and all were linked to a network of routes along the coast (see the map on page 171) and across the oceans. Larger ones such as Cape Coast Castle on the Gulf of Guinea also acted as headquarters for the British presence in the region.

Watson began his visit at Fort Gambia on James Island (named after the Duke of York, later King James II, and now known as Kunta Kinteh Island). His report describes 'a small irregular Square upon a small Island about ten leagues up the River Gambia', which 'if it was in good Order, would command the River very well'. The fort's strategic position in the lower reaches of this major river made it one of the most important fortifications in West Africa. It was also perfectly placed both to receive inland trade, and to act as a forwarding station for Caribbean-bound ships. Originally its main trade was in gold, but this had switched to slaves by the time this map was made.

In the top left corner stood the quarters and yards where slaves were kept while awaiting transport overseas. These contrasted with the 'slave holes' or dungeon holding areas found in other forts. At lower right, the landing place for ships has a boat house, cooperage and stores for pitch tar nearby. This low island in a broad river was the last that many slaves saw of their homeland. Other slaves – lists of whom appear among the fort records – were forced to work on the island itself, and lived in the circular huts around the unwalled lower edge and in the lee of the fort.

Watson's detailed plan shows the central fort with defensive corner bastions, a tall watchtower from which to spy approaching ships, and rooms inside for the governor, council and officers. Along the island's defensive outer walls lay barracks, stores and services: a smithy, saw pit, lime kiln, granary and kitchens. Watson notes several fallen roofs, and that 'through ... want of proper Repairs, it is all most gone to Ruins'. His report resulted in a House of Commons grant to carry out the required work, but within 25 years the fort was destroyed by the French, and within 50 years the slave trade was abolished in the British Empire. This plan remains to show the layout of a long-gone fortress, replicated in many similar forts along this coast, and to give an insight into the role played by the slave trade in European colonial expansion.

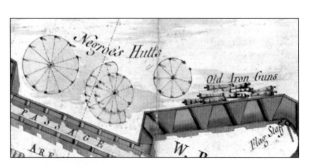

FLAG FORT: The flagstaff was an important part of any British fort. This detail also shows two designs of hut, perhaps in grass, alongside some 'old iron guns'.

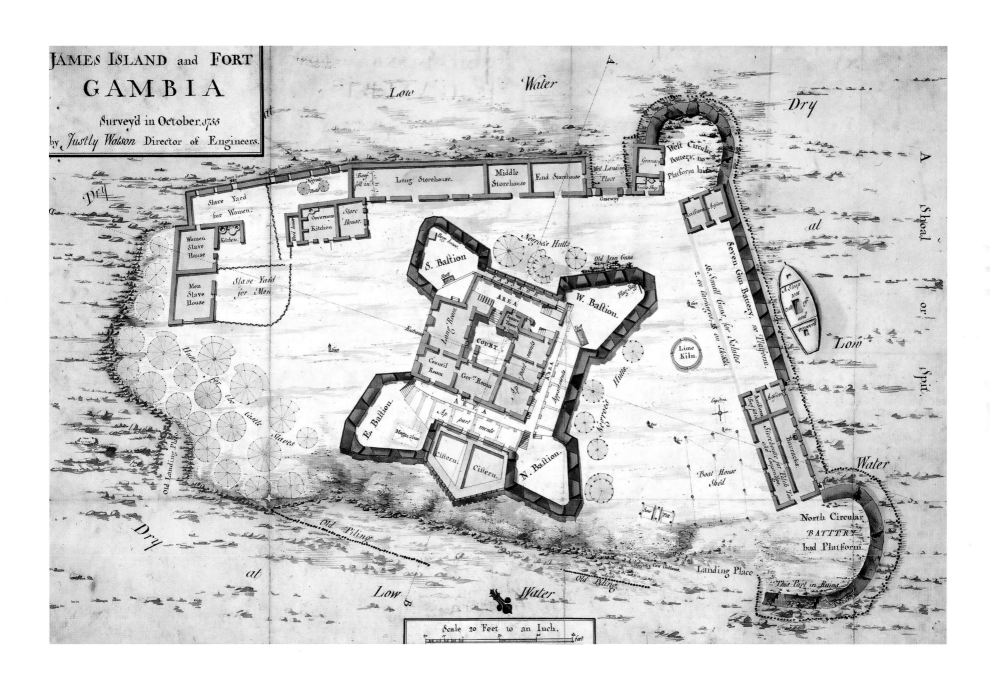

JAMES ISLAND and FORT
GAMBIA

Surveyd in October, 1755
by Justly Watson Director of Engineers.

Low          Water                    Dry

Dry

West Circular
Battery, no
Platform laid

Granary

Gateway

West Landing
Place

Roof
fell in.          Long Storehouse.          Middle
Storehouse.          End Storehouse

A Shoal

Slave Yard
for Women.

Negroe's
Hutts

Wood house

Governors
Kitchen          Store
House.

Kitchen

Women
Slave
House

Men
Slave
House          Slave Yard
for Men.

Bow house          Negroe's Hutts.          Old Iron Guns

S. Bastion          AREA          W. Bastion.          Flag Staff

Square
Tower          Seven Gun Battery, no Platform.

Entrance          Long Room          COURT.          Apartments

Council
Room          Gov.rs Room          Apartments          Lime
Kiln.

A Sloop
built &
repaird

Low

Spit

Store house for Wild Beast

Barracks.

AREA          Apartments

Hutts          Negroe's Hutts.

E. Bastion.          Ap      part   ments

Magazine          N. Bastion.          Boat House
Shed

Cistern          Cistern.

Hutts   for   the   Castle   Slaves          Water

Saw Pit

Old Landing Place          North Circular
BATTERY
bad Platform.

Dry          Old  Piling

at          Landing Place          This Part in Ruins.

Low          Water          Old Piling

Scale 20 Feet to an Inch.

# Rescued by Indians

The story has it that William Bonar, who made this map, was sent to the French Fort Toulouse to find out useful military information but, although disguised as a pack-horseman, he was caught and arrested as a spy. Simply labelled 'French fort' on the map at a fork in the river to the left of centre, Fort Toulouse had been built by the French earlier in the 18th century at a natural meeting point of rivers and Indian nations, to try to counter the growing influence of the neighbouring British colonies of Georgia and Carolina. At the point when this map was drawn, France and Britain were opposed in the Seven Years War.

The middle drawing to the right of the map is a building shown in plan, which may be an attempt by Bonar to record Fort Toulouse as noted during his captivity there. Bonar was rescued by a party of pro-English Creek Indians (or Muscogee), at the request of South Carolina's governor, Samuel Pepper, to whom Bonar was acting as an aide. Bonar would later revisit the fort as a lieutenant in the provincial militia, when this intelligence from his earlier reconnoitre was no doubt very useful.

This map with its surrounding pictures showing details of Creek life was probably drawn while Bonar was in Creek hands. The central sketch map has west at the top, and shows the heartland of the Creek Indians (to the south of those of the Cherokee),

between the Coosa River and the Chattahoochee River, with the Mobile River at left leading off towards the ocean. There is detail of the many named creeks off these rivers. The map shows the division of peoples among Upper Creeks and Lower Creeks, according to whether they used Lower, Middle or Upper Trading Paths to connect them with South Carolina, indicated by dotted lines at the right of the map.

The drawings around the map offer a glimpse of some of the Creek people. Flanking the title, a male elder smoking a pipe faces a female elder. At lower right is a scene with a woman wearing a long dress and with ribbons in her hair, and a warrior carrying his spear, gun and bow, a quiver of arrows over his shoulder. More weapons are drawn at top right: tomahawks, clubs, swords, lances and guns, along with a drum. At top left is a 'hott house', which was a winter meeting house. Below it is the public square, a ceremonial and social space where meetings were held outdoors in the summer. The picture below shows a games field where 'chunkey' was played, using special sticks shown crossed on the ground to point out the player's guess as to where a disc would land. While the map is a useful indicator of the state of English knowledge of this area at that time, the drawings around it give us a rare and vivid picture of life in a native American village, as seen through an English colonist's eyes.

GATHERING PLACE: The public square, shown here set round a central sacred fire kept burning, was a general meeting place for the Creek people, each depicted in detail here, as though their upper bodies are silhouetted against the light.

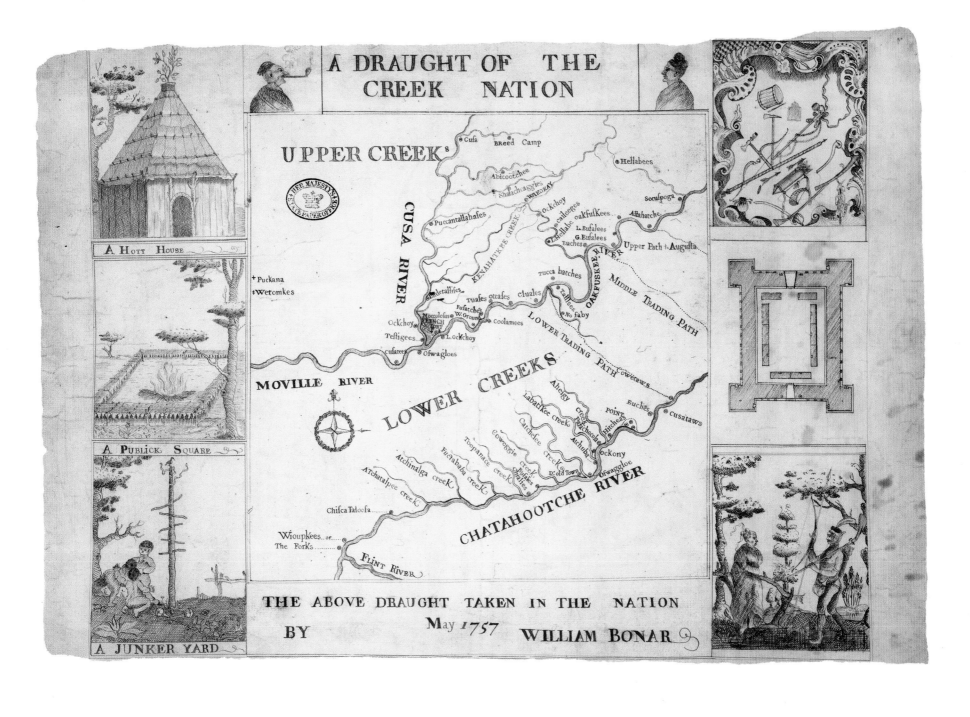

# A DRAUGHT OF THE CREEK NATION

A HOTT HOUSE

A PUBLICK SQUARE

A JUNKER YARD

UPPER CREEKs

Cufa · · BReed Camp

· Abicootchee

· Shalachaiggies

WAKOKAY

Puccantallahafes ·

CUSA RIVER

Ockchoy
calleeges
· oakfulkees
Latellafee
L. Eufalees
G. Eufalees
Buches

· Hellabees

· Soculpoga

Allahatche

KENAHATKPEE CREEK

Upper Path + Augusta

metallifes
Tuafes
otrafes
cluales

Tallifgee
No faby

tucca batches

OAKFUSKEE RIVER

MIDDLE TRADING PATH

Ockchoy
Fufatche
Moccolofus W. Ground
FRENCH FORT
Teftigees
L. ockchoy
culatee
Ofwagloes
coolamees

LOWER TRADING PATH

oweraws

MOVILLE RIVER

LOWER CREEKs

Ahelfy creek
Labackkee creek
Patchooth
POINT
Buches
Hitcheas
· cufataws

Catchelee
creeks
Adthube
ockony

Toopanate creek
Cowoggle
creek

Euchabatta creek

Atchinalga creek

D: old Town

Ofwaggeloe

ATchatalpee creek

CHATAHOOTCHE RIVER

Chifca Taloofa

Wioupkees or
The Forks

FLINT RIVER

+ Puckana
+ Wetomkes

THE ABOVE DRAUGHT TAKEN IN THE NATION
BY   May 1757   WILLIAM BONAR

# Wildlife on the border LAKE CHAMPLAIN, 1767

This is the top part of a map that shows an area on the border between the provinces of Quebec, New York and Vermont. The boundary of the first two is shown by a solid line dividing 'G' and 'H' at top right, while that between the latter two runs through the left side of Lake Champlain. A strategic waterway in a rugged land without viable roads, the lake and its tributaries gave access to the interior, and passage from the northeast to the Hudson River in the south. In the first half of the 18th century this land, home to a number of native American peoples, was hotly disputed between the French and the British. France's defeat by Great Britain at Quebec in 1759 and the Treaty of Paris in 1763 effectively ended French hopes of dominance in this region.

From the beginning of the century settlers from both countries had been given land grants, resulting in the complex pattern of landholding reflected in this map of 1767. The French claims 'express'd by dotted black Lines' around their square plots show their extensive settlement before that time; a table on the lower half of the map names 19 French seigneurs, keyed by letter to their land. Red lines denote 'Grants made to the English reduc'd Officers and disbanded Soldiers', and claims by other British settlers. The red and black lines overlap in places, showing how grants by the different colonial powers conflicted on the ground. As New York's

Governor Sir Henry Moore said of the map, 'it shews how far the Claims interfere with each other'.

Some insight into the map's genesis is given in Moore's letter to the Colonial Office of February 1767. He recalls that when writing his previous letter he had 'just received from Canada the Map of the French Settlements which I had so long expected'. The ship taking that letter was about to depart, so he had no time at that point to 'get the Claims of the Reduced Officers and disbanded Soldiers laid down upon it. This is now done on a reduced Scale to make it more portable', thus enabling the conflict between the claims to be appreciated 'at One View'.

The man who made this 'reduced Scale' map was Simon Metcalfe, Surveyor General of New York province. A Yorkshireman by birth, he drew these land grants in the manner of English estate maps, with neat plot lines, tree symbols, and a reference table, all suggestive of an ordered landscape. The title cartouche, too, recalls those on maps of his homeland, but instead of bucolic pastoral scenes, here a wild menagerie is gathered around a rugged rock face. Amid wolves, a crane and a turkey, the lion appears to have strayed from Africa, rather than down from the mountains. This contrast between the map and its decoration conveyed to those in London both the inherent possibilities and the dangers present in the colonies.

SCALING A TREE: This scene hides a novel scale statement. The standing child holds a pair of dividers against the tree, the trunk of which is marked in divisions corresponding to the scale of 1 inch to 4 miles. This is also written in words across the lower left hand leaf.

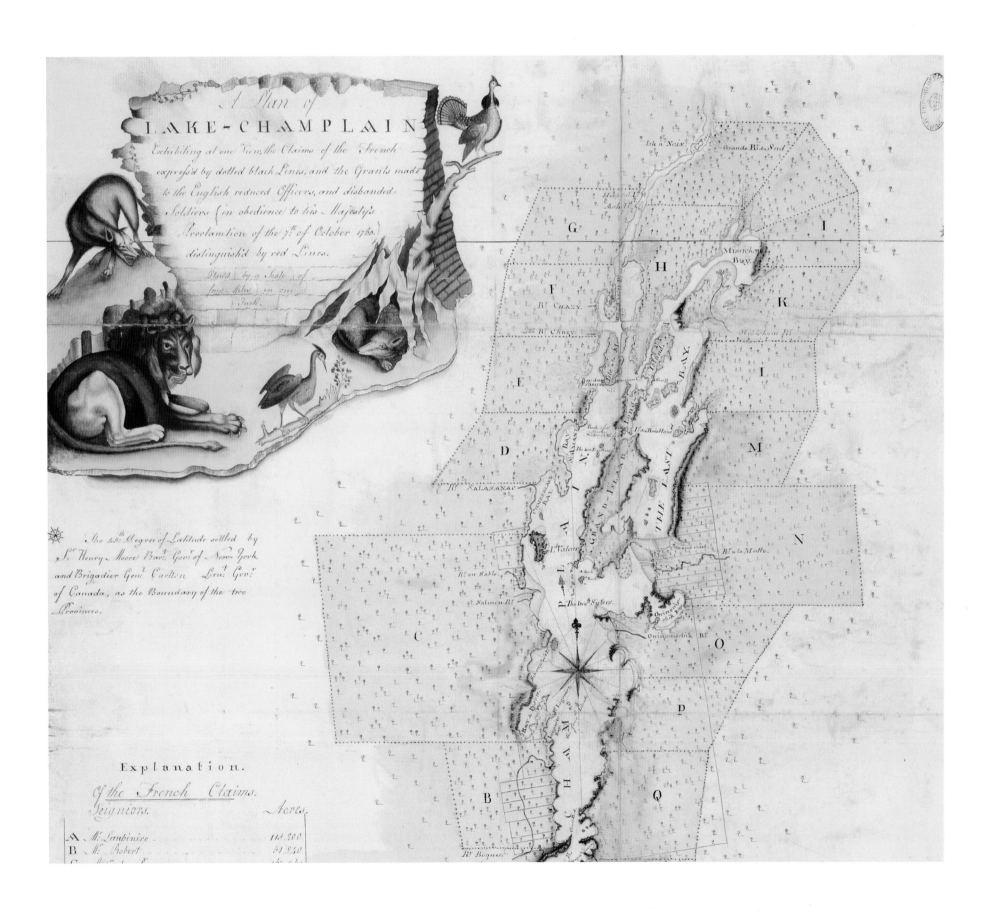

# Kidnap, kauri trees and the underworld

This is the oldest known Maori map of New Zealand. It was made because of a mistaken idea about who would know how to dress flax. About 400 miles away on Norfolk Island a type of this plant flourished, fanning the Royal Navy's hopes of self-sufficiency in ropes and sail-cloth. Nobody in the new British colony there knew how to make it usable, so the island's Governor, Philip Gidley King, arranged the kidnap of two people from New Zealand, where the same flax also grew. Unfortunately for him, his captives, Tuki Tahua and Huru, were high-status men who knew little about such women's skills. During their enforced stay, King made a Maori vocabulary and gazetteer which he sent to the Colonial Office with a letter in which he described the men, their language and their customs.

Also forwarded with the letter – as 'proof of this Man's abilities' – was this map made by 'Tooka-Titter-anue Wari-Ledo, a priest of that country'. Tuki originally drew it in chalk on the floor and transferred it to paper using a pencil. King then inked over the outlines, and annotated it. The Governor compared it favourably with Captain Cook's chart of New Zealand – a scientific coastal survey, with few interior features. Tuki's accomplishment is of a different kind from Cook's: it communicates his mental map of his homeland and records what was important to him. Within the framework of physical place, he evokes different dimensions of experience: social, political and mythical.

West is at the top of the map. North Island (at right) where Tuki lived is drawn larger than South Island, which he had never visited but about which he had heard; the wheel symbol within the latter indicates a lake with good stone for hatchets. No other lakes are shown, nor the mountains and geysers that we now associate with New Zealand. The map is concerned less with topography – or with mathematical accuracy – than with the networks and legends that were important from a Maori perspective. The notes about chiefs and clans describe who lived where, with whom they were feuding and to whom they were related. Trees left of top centre represent a kauri forest near a clan friendly with Tuki's people. At the far right, marked by a sacred tree, is Cape Reinga where spirits entered the underworld after death, having travelled the mythical pathway marked by the double-dotted line across North Island.

As a map that crosses between the temporal and spiritual realms, this is not drawn to scale. The Maori measured earthly journeys not by space but by time. The travelling distance between Tuki's home on Doubtless Bay and Huru's at the Bay of Islands was two days by land and one by canoe. Their voyage from Norfolk Island would take only twelve days by Royal Navy ship when eventually, after six months, they were returned home. This map remains as tangible evidence of an encounter between Maori and a British official, 50 years before organised European settlement of New Zealand.

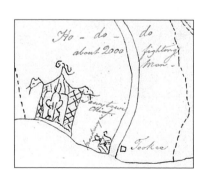

TUKI'S HOME: The large carved meeting house labelled 'Tewytewi' is near the *pa* or fort of chief Mudi Wai, and across the Oruru River from a small square symbol marked 'Tookee'.

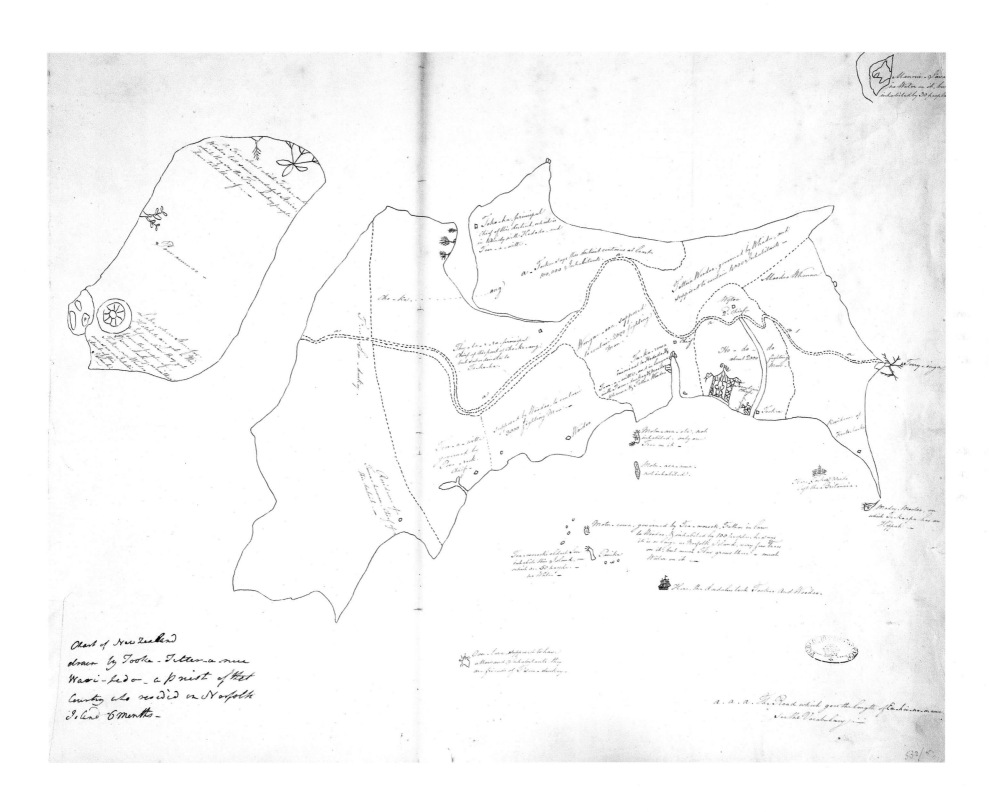

Chart of New Zealand
drawn by Tooka-Titter-a-nue
Warri-fedo - a Priest of that
Country who resided on Norfolk
Island 6 months -

# Kangaroo and campfire

This is the first known Aboriginal map drawn in pen and ink on paper. There is no indication of exactly where this is, or from which direction Galiput – its creator – intended it to be viewed, and it may not have mattered. It is accessible to us now, because Galiput explained its meaning to an onlooker, John Morgan, who recorded the legend for us in text below the map – or perhaps, to one side of it. This is written in a sort of pidgin English, as if trying to mimic a translator's pronunciation. Morgan also numbered key points important in the story of this 'native encampment' and its landscape.

Within the camp are separate areas for the women and children (at point 1), for the married men (written as point 7 but more likely to be point 4) while the single men camped some distance apart (point 2, at top right). Having set the scene of his tale, Galiput starts the action. 'Some morning sun get up vera early', when the married men would call the single men and all would go down to a nearby lake (point 6) to fish. Afterwards, the party would move on to hunt at a place where kangaroos were known to graze (point 5); one of the shapes here makes a plausible kangaroo, as rendered by someone unaccustomed to wielding a pen. The men would return to the campsite (point 3) to roast their catch on the fire, and all would sit round to enjoy the feast.

Galiput was from King George Sound, on the western south coast of Australia, and the map presumably shows his own home. He drew it while staying with Morgan, a minor official of the newly-founded Swan River Colony, near present-day Perth. Morgan had just taken up post and – so we learn from a letter that he sent to the Colonial Office with this map – he hoped that Galiput and another man called Manyat would improve communications with local Aborigines, which were not good. The two visitors had stayed with him for some time, and he records that they had gone to church on Sunday, and eaten a number of meals with him.

While Morgan wrote his letter, Galiput sat nearby 'amusing himself with a pen', and drew this map. Morgan was effectively witnessing how the Aborigine at his writing table – after a while of trying out the novelty of a pen, 'which he now holds tolerably well' – could readily adopt this new technology to communicate his view of his world. This interaction captured a long tradition of aboriginal mapmaking, where a stick or finger was used in sand or mud to make a sketch to narrate an event or to show how physical features related to one another. These maps in mud were smoothed over by the tide, and the sketches in sand blew away in the wind. Galiput's temporary access to the more permanent medium of ink on paper has allowed his creation to survive in the archives.

1955 Swan River

Galliput.

Particulars, [as expressed by Galipeet,] of the native encampment, scrawled out by him. —

That place [No 1] womanar, children, pickninny. — That place [No 7] — Married Men — that place [No 2] — Single Men —
Some morning sun get up vera early — married go down — call up single men, — single men get up when sun get up
very early. — all go down [to No 6 a lake] catch fish, then go up [to No 5] catch kangaroo. — bring him down
dare — [No 3] fire — roast him. — all men set round so. — [suiting the action to the word] upon ham.

# 'No white man had ever traversed the country before'

This map, with its tracing cloth blotched by moisture and ink runs, was drawn in the African interior. In a letter which accompanied the map the explorer David Livingstone described conditions there as 'broiling hot' – 96°F in the shade – and he noted that 'the air seems sultry and oppressive'. At the time of writing this letter Livingstone was leader of the Zambezi Expedition (1858–1864). This was funded by the Foreign Office, which had appointed him as honorary Consul for Quelimane, a port on the east coast shown diagonally across the lower edge of the map. David states that the sketch map, was drawn by 'Mr C. Livingstone' – his younger brother Charles, who was also on the expedition – although in fact it also bears annotations by David, who commended it for its 'accuracy and freshness'.

The aim of the map was to give an impression of the party's route from the coast, showing the rivers, lakes and terrain encountered, and especially to point out what Livingstone claimed as an 'English discovery' (i.e. by his party), shown by the blue line running from the four mouths of the Zambezi River at its delta, northwards to the shores of Lake Nyasa (Lake Malawi). This is not simply a route map. It also gives information about features and people encountered: the countries of the Zulu, Maravi and Manganja peoples, caravan routes, and a trade route to the coast along the Zambezi used by the Portuguese, who had been in east Africa for centuries.

Livingstone, in his enthusiasm to present his discoveries in a good light, made little of the natural obstacles he found on the way. A pointer to the place where 'navigation stopped', to the south-west of Lake Shirwa (or Chilwa), alludes to the fact that the party had to abandon boats as they encountered 'only 33 miles of cataracts' on their way up the River Shire, a northern tributary of the Zambezi. Indeed it was from Murchison's Cataracts that Livingstone addressed his letter to the Foreign Office on 15 October 1859. The letter travelled with the map out of Africa on the gunship HMS *Lynx* to reach London four and a half months later, on the 28 February 1860.

'Discovery of Lake Nyassa' was written on the back of the letter on receipt in London, but Livingstone, aware of his official position and funding, also wrote about other subjects he knew were of interest to the Foreign Secretary, Lord John Russell. The development of trade in commodities was seen as a way to supplant the local slave trade then still much in evidence. There is an optimistic tone in Livingstone's portrayal of the area when he states that 'We have opened a cotton and sugar producing country … that really seems to afford reasonable prospects of great commercial benefits to our country'. At the end of his letter Livingstone says 'We had no difficulties with the natives though ... no white man had traversed the country before'.

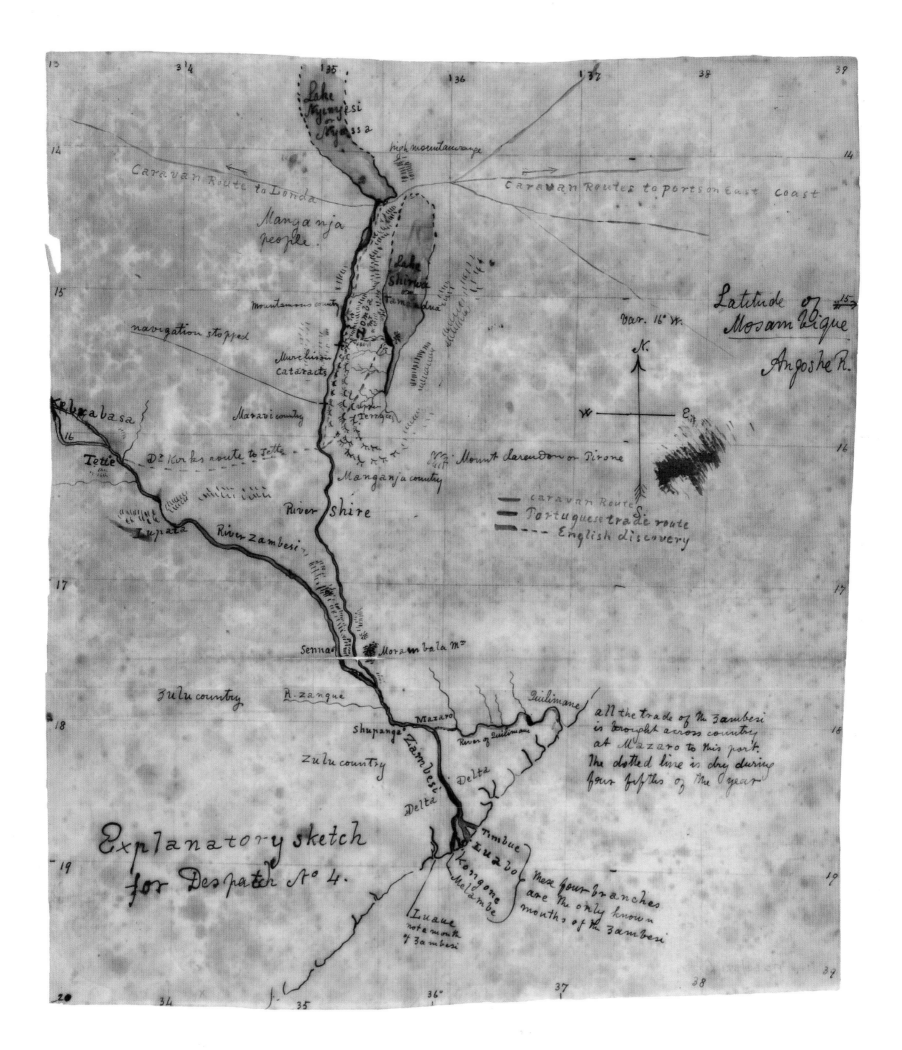

Lake
Nyinyesi
or
Nyassa

high mountain range

Caravan Route to Londa

Caravan Routes to ports on East Coast

Manganja
people.

Lake
Shirwa
Tamandua

mountainous country

navigation stopped

Murchison
Cataracts

Maravi country

Manganja country

River Shire

Var. 16° W.

N.

W                    E.

Latitude of
Mosambique
Angoshe R.

Mount Clarendon or Pirone

Caravan Route
Portuguese trade route
English discovery

Kebrabasa

Tette

Dr Kirks route to Tette

Lupata      River Zambesi

Sennα      Moramba la Mⁿ

Zulu country      R. zanque      Quilimane

Mazaro

Shupanga      River of Quilimane

Zulu country      Delta

Zambesi

Delta

Timbue
Luabo
Kongone
Melambe

Luaue
not a mouth
of Zambesi

all the trade of the Zambesi
is brought across country
at Mazaro to this port.
The dotted line is dry during
four fifths of the year

These four branches
are the only known
mouths of the Zambesi.

Explanatory sketch
for Despatch N° 4.

# Journeys to the centre of the earth

In the late 1920s Britain still saw herself as the centre of a vast empire of countries interconnected by strong links of trade and communications, as shown by this map-poster. 'Highways of Empire' was not a map to navigate by, nor an example of precision cartography, but a piece of pure propaganda produced for the Empire Marketing Board. The Board was set up under the Dominions Office in May 1926 to encourage people in Britain to buy Empire produce, particularly foodstuffs. The assumption was that such sales would increase the purchasing power of British colonies, increasing exports of British manufactured goods.

A major poster campaign was envisaged as a key element in this marketing strategy. It was decided to launch the campaign with a special map which would emphasise Britain's role as the hub of the Empire. Files in The National Archives show how the Posters Subcommittee realised that a map displayed on a 48-sheet poster measuring 20 feet by 10 feet would only magnify the distortions inherent in the usual Mercator's projection on which most world maps were made. The Committee sketched an idea for a completely new projection for this poster, which would place the British Isles at the centre of the map, within a half moon to suggest the globe. They commissioned the cartographic firm George Philip and Son to draw the outline.

The man they asked to design the map on this new base was artist MacDonald Gill, already well known for his classic 'Wonderground' map for the London Underground, which was striking for a combination of geographical basis with quirky details based on legends, famous sights and quotations.

The result was unveiled in a blaze of publicity on New Year's Day 1927, when police had to control crowds around the hoarding on which the poster was displayed in Charing Cross Road. The map shows the world, British colonies and dominions suitably coloured bright red, with major shipping routes and banners expressing patriotic sentiments. MacDonald Gill framed the map by personified winds which seemingly blow ships along the highways of the British empire, upon which sun and moon figures suggest that the sun never set. These elements reminiscent of old classical world maps connote a continuity of power, while eye-catching colours and modern elements such as steamships and a bi-plane on the map itself emphasise contemporary commerce.

Gill was paid £150 for this design – a very considerable sum at that date. Around 3,000 copies of the 48-sheet version of the map were printed. It was also produced in two smaller versions for sale to the general public, and a miniature version was distributed at the Schoolboy's Own Exhibition in 1929.

UP THE WRONG POLE: Among the many decorative features on the map are polar bears. MacDonald Gill mistakenly drew them at both poles, when they only live at the North Pole. This error was only spotted at the proofs stage, when it would have been expensive to correct by complete erasure. A Committee member suggested that the situation might be redeemed by turning the polar bears into a comical feature, and speech bubbles were duly added to the bears, so one at the South Pole demands to know 'Why are we here?', and another sings 'It's a long way to Tipperary'!

# HIGHWAYS OF EMPIRE

# BUY EMPIRE GOODS FROM HOME AND OVERSEAS

R.S.W.B.O. **ISSUED BY THE EMPIRE MARKETING BOARD.**

PRINTED FOR HIS MAJESTY'S STATIONERY OFFICE BY. JORDISON & Cº LTD LONDON & MIDDLESBROUGH.

# East of Aden INDIAN SUBCONTINENT, 1948

When this map was made in the early summer of 1948 Indian independence was less than a year old. Mahatma Gandhi had died earlier in that year, and the last viceroy, Lord Mountbatten, had finally departed just a few weeks before, after a handover of his palace. The new political pattern was still not completely formed, and this map tells part of that story. The apparent simplicity of its design belies a complex situation, both on the map and on the ground. A range of historical, religious and territorial factors meant that the area of India previously administered by the British became two states – India and Pakistan – but three land masses. The western and eastern portions of Pakistan were separated by 1,000 miles. The human and administrative difficulties of this transfer of power were enormous.

There were over 500 Princely States in this area that had enjoyed a quasi-independent status under British rule. At partition, each was given a choice: to join India or Pakistan (which most did) or to remain independent. Hyderabad, the largest and most prosperous of these states, had been ruled from 1724 by an hereditary prince, the Nizam. As the map shows, its location was in the centre of the new Union of India, and it was this that would doom its continued separate identity. The Muslim Nizam, head of a mainly Hindu population, aimed to keep his domain but, aware of the threat from Hindu India, appealed for external help.

This map was made to brief the Commonwealth Affairs Committee of the British Cabinet on the whole question of the remaining Princely States, especially Hyderabad. It gives an insight into the work of the Foreign Office Research Department in preparing briefing documents to assist officials in their deliberations. Maps were made in-house, in international boundary cases or complex political situations such as this one, where they swiftly conveyed the essentials of a situation.

The map shows diagrammatically the new Dominions of India (left uncoloured) and of Pakistan (shaded green), with Hyderabad and the few other remaining Princely States coloured orange-yellow to highlight their position. This makes it immediately obvious how large and central an area Hyderabad occupied within India.

The map was filed with Foreign Office correspondence on the subject of the Princely States over the summer and into the autumn of 1948. A wide range of factors and opinions are discussed, especially the international aspect to this dispute between Hyderabad and India. One concern was that any attempt at annexation by India might prompt retaliatory action by Muslim Pakistan to help the Nizam, potentially sparking war. After the death of Pakistan's first Governor-General Muhammad Ali Jinnah on 11 September, events moved swiftly. Within a short time the Indian Army invaded Hyderabad, which – being completely surrounded – surrendered. This map is thus one of the last to show Hyderabad as an independent state before it became part of the new India.

LAST DAYS OF THE RAJ: Cecil Beaton's series of photographs for the wartime Ministry of Information capture the interior of the splendid Viceroy's Palace in 1944, at the end of British rule.

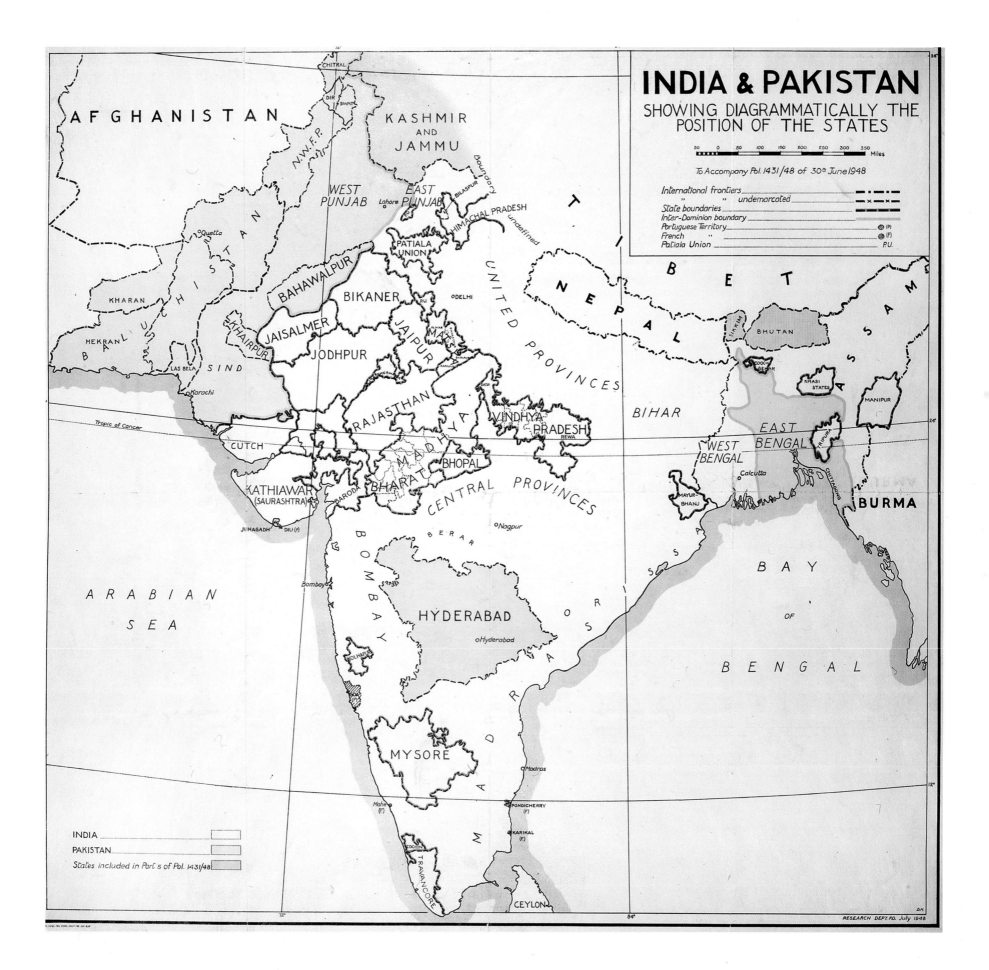

# INDIA & PAKISTAN
## SHOWING DIAGRAMMATICALLY THE POSITION OF THE STATES

To Accompany Pol. 1431/48 of 30ᵗʰ June 1948

50  0  50  100  150  200  250  300  350 Miles

International frontiers ................. —·—·—·—
  "   "   undemarcated ................. —×—×—×—
State boundaries ................. _____
Inter-Dominion boundary ................. _____
Portuguese Territory ................. (P)
French   "   ................. (F)
Patiala Union ................. P.U.

AFGHANISTAN

KASHMIR AND JAMMU

N.W.F.P.

CHITRAL

DIR  SWAT

WEST PUNJAB

EAST PUNJAB

Lahore

Boundary undefined

HIMACHAL PRADESH

BILASPUR

TIBET

NEPAL

BHUTAN

SIKKIM

COOCH BEHAR

ASSAM

KHASI STATES

MANIPUR

oQuetta

BALUCHISTAN

KHARAN

MEKRAN

LAS BELA

SIND

KHAIRPUR

oKarachi

BAHAWALPUR

BIKANER

JAISALMER

JODHPUR

JAIPUR

PATIALA UNION

P.U.

oDELHI

ALWAR

UNITED PROVINCES

BIHAR

WEST BENGAL

EAST BENGAL

Calcutta

CHITTAGONG

BURMA

TRIPURA

Tropic of Cancer

CUTCH

RAJASTHAN

VINDHYA PRADESH

REWA

DATIA

MADHYA BHARAT

BHOPAL

KATHIAWAR (SAURASHTRA)

BARODA

JUNAGADH  DIU (P)

CENTRAL PROVINCES

BERAR

oNagpur

MAYUR-BHANJ

ORISSA

ARABIAN SEA

Bombayo

BOMBAY

KOLHAPUR

GOA

HYDERABAD

oHyderabad

BAY OF BENGAL

MYSORE

MADRAS

oMadras

Maheo (F)

PONDICHERRY (F)

KARIKAL (F)

COCHIN

TRAVANCORE

CEYLON

INDIA
PAKISTAN
States included in Part 5 of Pol. 1431/48

RESEARCH DEPT. F.O. July 1948

# CHAPTER SEVEN

# MAPS THAT WITNESSED history

The world is always changing, both in subtle and more conspicuous ways. Encompassing everything from the natural cycle of life and death to the latest technological inventions, the essence and scope of these changes are extraordinarily diverse. Some are rapid and others gradual. Some have a far-reaching impact whilst others pass unnoticed. Some alterations to the physical environment, such as the construction of a new road, are planned by human beings; others, like the movement of tectonic plates beneath the earth's surface, occur independently of human activity. The one constant of today's world is that it is never completely identical to that of yesterday.

Maps, then, must capture places in time as well as in space. Yet the relationship between maps and time is not always a simple one. There is a saying that every map is out of date: the area depicted will always have changed to some degree since the map was created. Published mapping based on formal topographical surveys will be out of date

even before it is printed, as soon as the surveyors complete their work in the field. Many maps draw on the work of original surveys and revisions carried out over months, years or even decades. The discrepancies caused by such time lags will generally be insignificant to someone using a map when it is new. Conversely, older maps are often valued precisely because they provide evidence of the otherwise 'lost' landscapes of years past. Certain maps are deliberately historical, intended to show places in the past rather than the present, and some – such as those of Port Royal and Imperial Russia on pages 199 and 201 – are designed to depict changes over time.

This seventh chapter has a different focus from the preceding six. Rather than concentrating on a particular type of map or purpose for using maps, it follows the theme of moments in time and how maps reflect historical change. A number of our chosen maps record situations of evident global importance. Others may appear to be of more

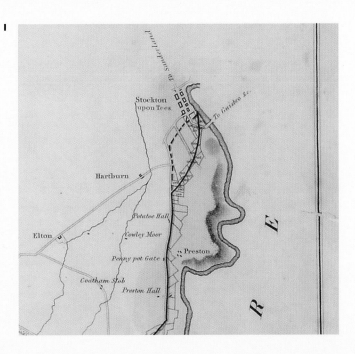

**1** INDUSTRIAL REVOLUTION: At the end of the 18th century, Stockton-on-Tees grew from a small market town into a centre of heavy industry. In 1825, it became the terminus of the world's first permanent passenger railway (see page 205).

**2** A NEW NATION: Great Britain acknowledged the independence of the United States of America in 1783. The following year, Abel Buell produced the first published map of the new country to be created wholly on American soil.

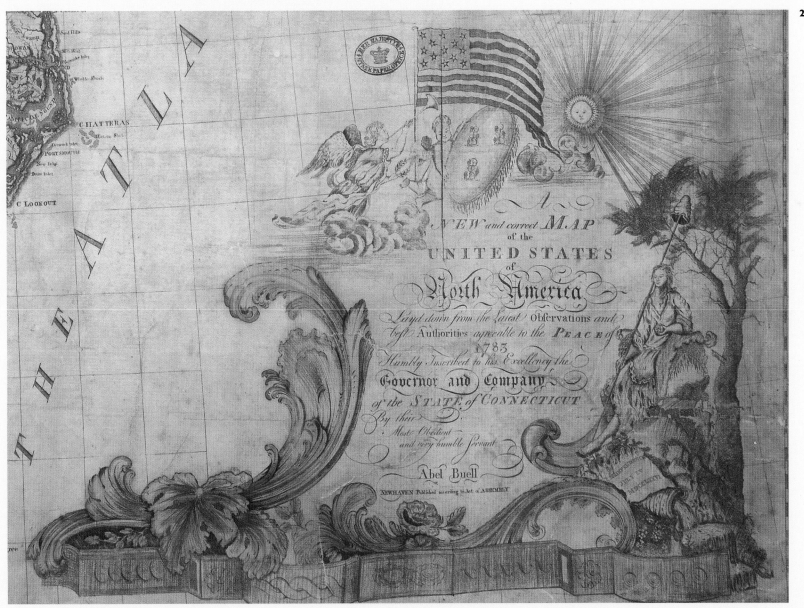

localised interest, but gain greater significance when set within their broader context. What our selected maps have in common is that each reflects a decisive event, a significant moment, or a particular perspective on a situation in history. The importance of the moment may extend beyond the immediately obvious. For instance, the annotations on the map of Czechoslovakia on page 215 record not only an external event (Adolf Hitler's demand for territory) but also a pivotal step in the internal history of the map that we have today: its transformation from a mass-produced item to a unique historic document.

The content of this chapter is not, however, sharply distinct from earlier portions of this book. We have already explored a broad spectrum of noteworthy events, from the birth of the city of Adelaide (see page 67) to the partition of India upon its independence from British rule (page 191), and from the natural eruption of Vesuvius (page 147) to the human-wrought devastation of the London Blitz (page 75). Just as those maps looked forward to the theme of this chapter, here we also look back at earlier themes. Our 400-year journey takes us from a hand-drawn, pictorial 16th-century map (page 197) to the published work of the British Admiralty in the 1950s (page 219). On the way, we pass through urban and rural landscapes, confront scenes of conflict and the horrors of war, and witness the rise and fall of empires. In fact, since the contents of the archives are by definition survivors of the ravages of time – whether intentionally or by happy accident – we could easily have chosen to call the whole book 'Maps that witnessed history'.

What, then, is distinctive about the maps included in this chapter? The answer is a difference in emphasis, which lies not so much in the inherent qualities of the maps themselves as in the balance of our approach to them. The stories woven around the twelve maps featured here tend to concentrate less on the specific content and origin of each one than on the wider context of the historical events surrounding it. In these short narratives, cartography becomes a portal for exploring the world of the past.

**3** A THIN RED LINE: Robert Schomburgk's 1841 survey portrays a British claim for where the border between Venezuela and British Guiana (now Guyana) should lie. This map's influence has extended across time into subsequent negotiations over this boundary.

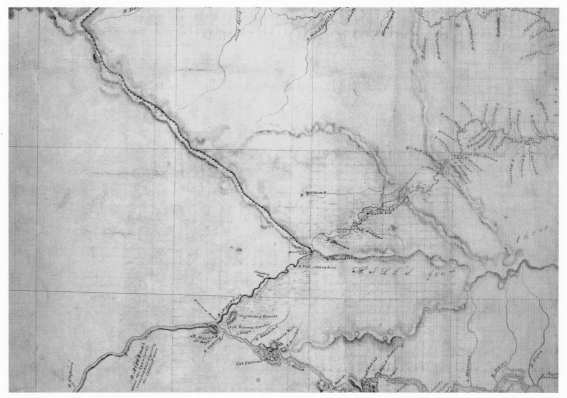

**4** EAST AND WEST: After the Second World War, the victorious allies divided the German capital of Berlin into the four zones shown here. The formal separation between East Berlin (the Soviet zone) and West Berlin (the other zones) lasted from 1949 until 1990.

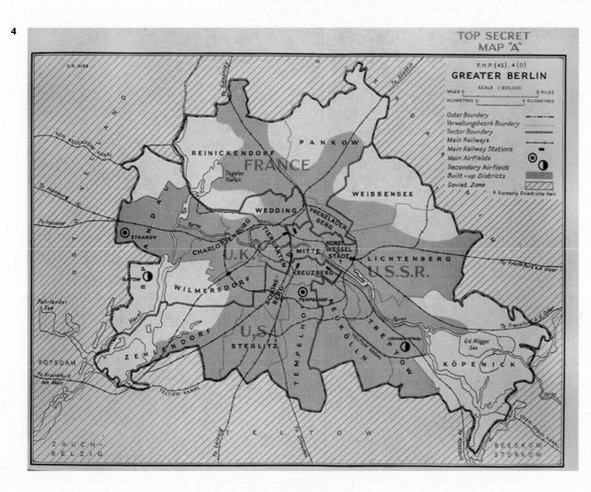

# 4

**5** SECRET AGREEMENT: In May 1916, the British and French governments signed a pact to divide a large portion of the Middle East between them if they won the First World War. Much of the Sykes-Picot line separating the French and British zones (labelled A and B, respectively, on this map) remains the basis of current borders.

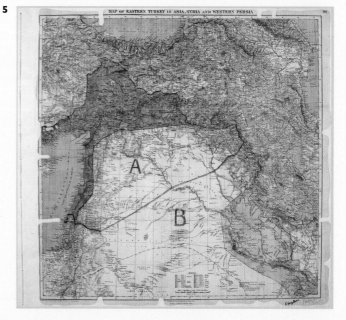

# 5

# A murder mystery

This rather gruesome item resembles a strip cartoon almost as much as it does a conventional map. It depicts the death and burial of Henry Stewart, Duke of Albany – commonly known as Lord Darnley – the second husband of Mary, Queen of Scots. Many details, particularly the human figures, are drawn out of proportion with one another, presenting a disconcerting impression to the modern viewer. The plan was enclosed in a letter from Sir William Drury, Marshall of Berwick, to William Cecil (afterwards Lord Burghley), who was chief adviser to Queen Elizabeth I of England, Mary's cousin and rival. Drury served as Cecil's eyes and ears in Scotland, keeping him informed of events likely to interest Elizabeth and her government.

Darnley was born in England of mixed Scottish and English ancestry. He was related to the royal families of both countries and his relatives were deeply involved in the political rivalry and diplomatic manoeuvres between the two nations and their sovereigns. When he married Mary in 1556, Darnley was 19; she was three years older and the widow of the King of France. Although Mary was initially very fond of her husband, their marriage was not happy. Less than eight months after the wedding Darnley was implicated in the violent killing of Mary's secretary, David Riccio, who was stabbed to death in front of the pregnant queen. Thereafter, they lived largely separate lives.

Early in the morning of 10 February 1567, the bodies of Darnley and his servant were found in the garden of the Old Provosts' House at Kirk O'Field, then on the outskirts of Edinburgh, the Scottish capital. This is shown in the upper right-hand part of the plan. Although a dagger was found lying on the ground nearby, the bodies bore no wounds or marks to indicate how the men had died; they may have been suffocated. The house itself had been destroyed by an explosion (shown left of the centre) during the night. The groups of figures in the lower part of the plan represent Darnley's corpse being carried away and buried. Nobody was ever convicted of killing the two men, although James Hepburn, the Earl of Bothwell – who became Mary's third husband soon afterwards – was widely blamed at the time. It is now thought that a group of Scottish noblemen, including Bothwell, planned Darnley's murder.

Darnley and Mary's baby son, James, is depicted on the upper left-hand side of the plan. Although he was not old enough to talk, artistic licence has been employed to portray him as offering a prayer for divine retribution against his father's killers. A few months after this plan was drawn, Mary was forced to abdicate in favour of James. After spending many years under house arrest in England, she was convicted of plotting against Elizabeth and executed in 1587. When Elizabeth died in 1603, King James VI of Scotland succeeded her as James I of England. Through him, Mary and Darnley are direct ancestors of the present British Royal Family.

SON AND HEIR: This scroll, reading 'Judge and revenge my caus[e], O Lord', serves as the 16th-century equivalent of a speech bubble for the infant Prince James.

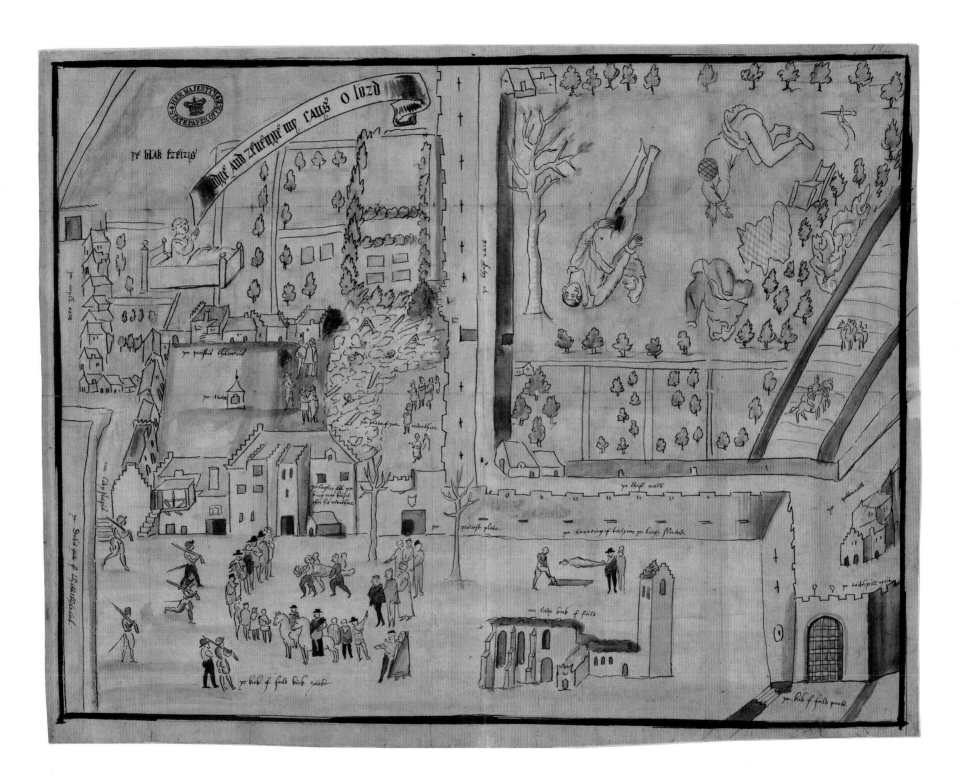

# Lost beneath the waves

PORT ROYAL, JAMAICA, 1692–1870

When the English captured Jamaica from the Spanish in 1655, the old capital of St Jago de la Vega (now known as Spanish Town) was badly damaged during the invasion. Under English rule, the new settlement of Port Royal nearby rapidly became the island's chief centre of commerce. Within a few decades it had become the richest city in the Caribbean. As a magnet for pirates and privateers, Port Royal gained a reputation as 'the wickedest town in Christendom'. After a powerful earthquake shook eastern Jamaica in 1692, destroying most of the city, some contemporary religious commentators proclaimed it a just punishment for the inhabitants' sins, comparable to the destruction of the biblical Sodom and Gomorrah. To our eyes, a more apt comparison is with the legendary Atlantis, a fabulously wealthy ancient city said to have been lost under the sea.

Port Royal enjoyed the economic advantages of a site at the entrance to a large natural harbour but the area was prone to tremors and strong winds. Much of the city had been constructed on unstable land and many of the closely-packed buildings lacked adequate foundations. When the earthquake struck shortly before noon on 7 June 1692, followed by a huge tidal wave, about two thirds of the city sank beneath the sea. The difference between the coastline before the earthquake (outlined on this map in blue) and afterwards (outlined in red) is a stark reminder of the power of nature over human endeavour.

Only one tenth of Port Royal's buildings survived the earthquake and many of these remaining homes and businesses were looted by the survivors. Although some of the city was later rebuilt, its recovery was hampered by several later disasters – including fires, floods, hurricanes and smaller earthquakes – and it never regained its former status. A new commercial centre was established in nearby Kingston, which eventually supplanted Spanish Town as the official capital of Jamaica. Its urban area has since spread to encompass Port Royal.

Although this map was not created until 1870, it has a close connection with the events of 1692. Thomas Harrison, a government surveyor, copied it from an earlier survey of 1827 and added the coloured outlines to indicate how the shape of the peninsula had altered. Suspecting that money, jewels and other precious objects believed to have been submerged during the earthquake might be recoverable using modern diving equipment, local officials debated whether they should organise their own diving expedition or allow a private venture. They referred the matter to the Colonial Office in London, enclosing this map with the correspondence. The British government agreed that the Governor of Jamaica could issue a licence to a firm called Stein & Company to undertake a search; the Crown waived its right to a share of any profits in favour of the Jamaican government. We know that Stein & Company accepted the terms because its director wrote to thank the Colonial Secretary for his support, but our records contain no evidence that divers found any sunken treasure.

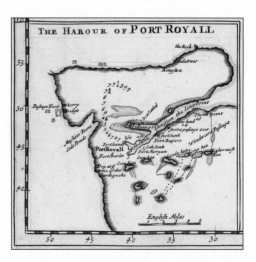

PIRATE CAPITAL: The harbour of Port Royal in about 1700, shortly after the time of the earthquake.

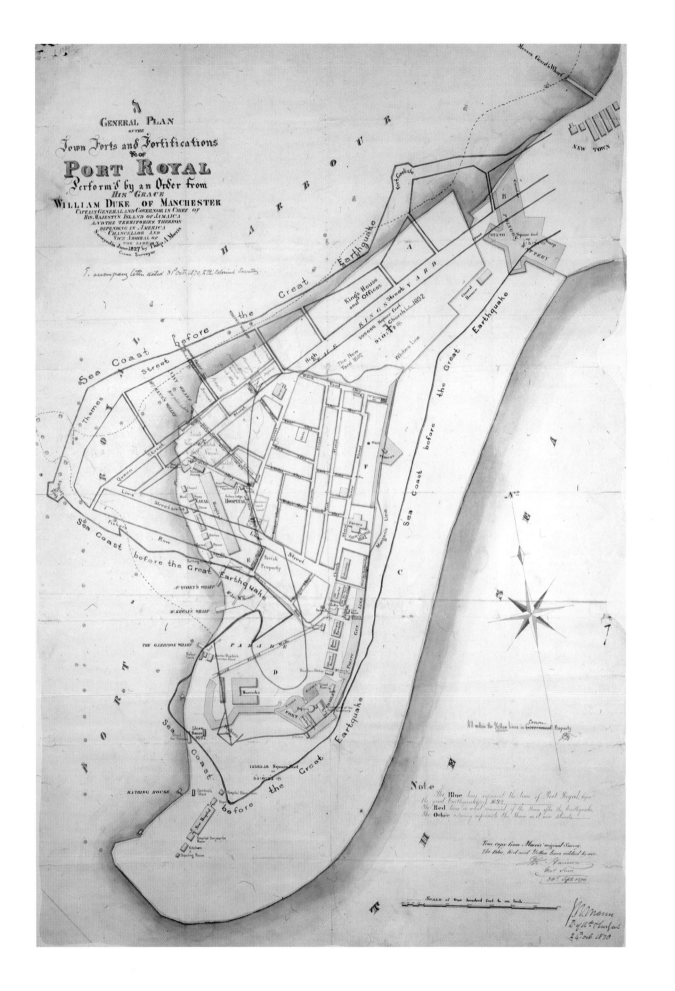

# Realm of the Romanovs   RUSSIA IN EUROPE, 1763–1836

For most of the 19th century the name Hertslet was synonymous with the Foreign Office library. Lewis Hertslet was appointed as sub-librarian in 1801 and promoted to chief librarian in 1810, overseeing both the department's collections of publications and its recordkeeping. On his retirement in 1857 his youngest son, Edward, succeeded him as head of the library, serving until 1896. Both father and son developed an encyclopaedic knowledge of political history and geography. By placing the management of information at the heart of government, the Hertslets supported the effective conduct of the United Kingdom's diplomatic relations and its interest in world affairs. It was Lewis Hertslet who produced this hand-drawn map showing the westward expansion of the Russian Empire between 1763 and 1836. Damage along and repairs to some of its edges and folds indicate that it was heavily used. The many amendments in pencil may be corrections prior to the making of an engraved copy but we know of no extant printed maps derived directly from this original manuscript.

In contrast to the maritime empires of overseas colonies built up by the western European powers, Russia had grown as a land-based, largely contiguous realm, spreading steadily eastward into Siberia and Central Asia. For a time, the empire spanned not just Europe and Asia but America too: from 1799 until its sale to the United States of America in 1867 Alaska was a Russian colony. Under the Romanov dynasty, which ruled between 1613 and 1917, Russia turned her attention to the south – where the great empires of Persia and Ottoman Turkey lay – and the west, as well as the east. The modernising Tsar Peter the Great (1682–1721) moved his capital from Moscow to the new city of St Petersburg in 1703 to provide his empire with a 'window upon Europe'.

The progressive territorial expansions to the west and southwest shown on this map were achieved through a mixture of warfare and diplomacy, particularly during the reigns of Catherine the Great (1762–1796) and Alexander I (1801–1825). The majority of Russia's new possessions were acquired from the fringes of the Ottoman Empire or at the expense of Poland, which was gradually partitioned among Russia, Prussia and Austria. Her single most substantial acquisition of territory, however, was the Grand Duchy of Finland, transferred from Sweden under the 1809 Treaty of Fredrikshamn.

Russia's military successes and growing political importance during the early 19th century – cemented by Tsar Alexander's victory against the French Emperor Napoleon in 1812 – made her of particular interest to the British government. In the context of the frequent wars and shifting alliances that characterised the European power politics of this period, it was crucial for the Foreign Office to maintain ready access to information about the United Kingdom's allies and rivals. Whilst the near-century of service from such high-calibre officials as the Hertslets played a pivotal role in the accumulation of such knowledge, maps of this kind proved to be equally valuable as a method of capturing and expressing it.

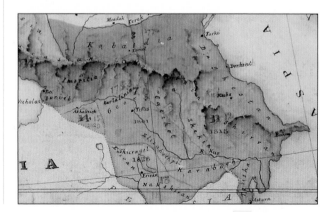

CAUCASUS: The territories of the Russian, Persian and Ottoman Empires converged in this ethnically and linguistically diverse region. During the late 18th and 19th centuries, Russian control of the Caucasus increased as the power of its southern neighbours waned.

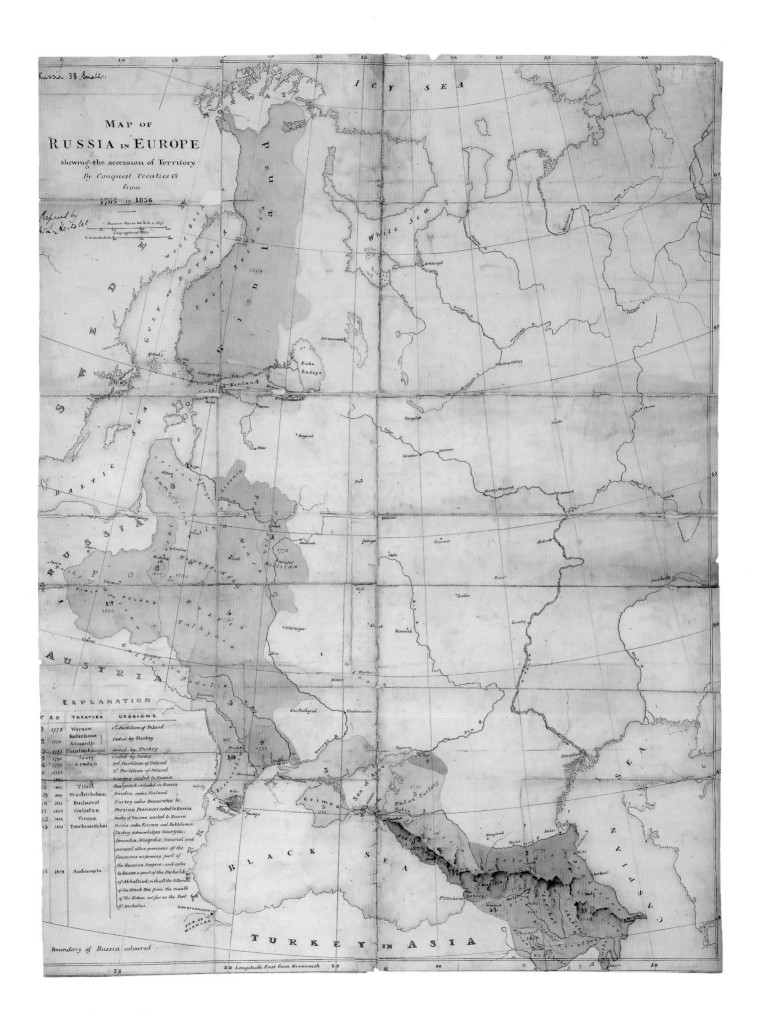

MAP OF
RUSSIA IN EUROPE
shewing the accession of Territory
By Conquest, Treaties &c
from
1763 to 1836

# An emperor in exile ST HELENA, 1815

This attractive printed map sheds light on the last years of one of history's most famous political and military leaders, Napoleon Bonaparte. Born in Corsica, he rose from obscurity after the French revolution to become a senior army officer. In 1799, aged 30, he led a successful military coup and established himself as the leader of France, initially as first consul and later as emperor. His reign was marked by a series of conflicts now known as the Napoleonic Wars. At first, France was in the ascendant and Napoleon came to dominate much of Europe, but eventually the tide turned against him. Comprehensively beaten by the Sixth Coalition of his many opponents, he was deposed and exiled to the Mediterranean island of Elba in 1814. Escaping from Elba the following year, he resumed the role of emperor for a few months but was finally defeated at the Battle of Waterloo and banished once more.

For his second place of exile, Napoleon's opponents chose one of the most remote places on Earth, the island of St Helena. Lying in the Atlantic Ocean almost 1,200 miles west of the coast of Africa and controlled by the British East India Company, the island was already fortified and relatively easy to defend. During Napoleon's exile, the British government took a direct role in the administration of St Helena. An army officer called Sir Hudson Lowe was appointed as the island's governor, reporting to the Secretary of State for War and the Colonies. Although Lowe

had been carefully chosen as a suitable guardian for the former emperor, relations between the two men were always fractious. The East India Company resumed full control of St Helena when Napoleon died in 1821 but relinquished it to the Colonial Office in 1834. It remains a British overseas territory today.

This map of St Helena was produced and sold as a souvenir commemorating the defeat of the United Kingdom's arch-enemy. Designed by Lieutenant R P Read and dedicated to the Duke of Kent (whose daughter later became Queen Victoria), it shows the island's rugged terrain, agricultural estates and main residences. The map is dissected and mounted on cloth so that it could readily be folded for storage. A note on the back of our copy reveals that it was formerly owned by Edward Vigers – a London-based surveyor and architect whose first wife, Flora Alexander, came from St Helena – and given to the Colonial Office on 18 August 1911.

The first edition of the map, dated 1815, was apparently rushed into print and contains a mistake. It labels Plantation House as the former emperor's residence and Longwood House as Governor Lowe's home. Later editions were corrected to show that the reverse was true. Napoleon would doubtless have preferred to live at Plantation House, for he disliked Longwood, complaining that it was damp and unhealthy. Now owned by the French government, the house has become a museum dedicated to St Helena's best known former resident.

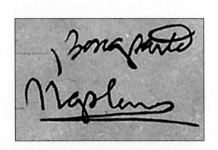

FAMOUS NAME: The map bears facsimiles of Napoleon's signatures as first consul and emperor.

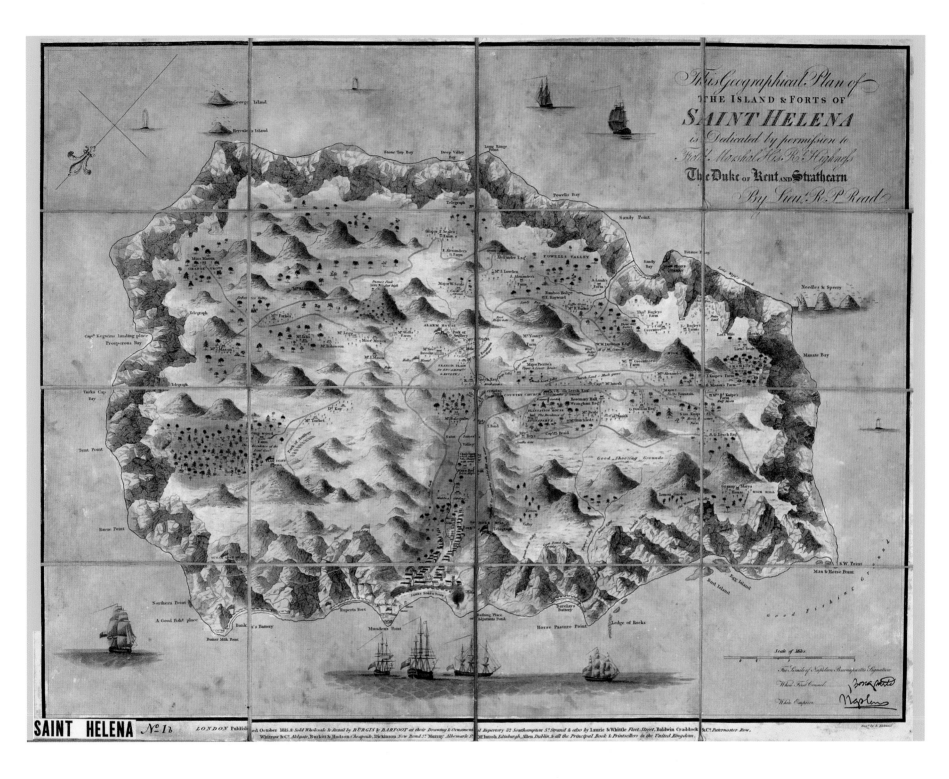

THE EMPEROR'S PRISON:

Longwood House is labelled

incorrectly on this first edition as the

governor's residence but correctly on

later editions as Napoleon's home.

# Steaming ahead

When a crowd gathered in Stockton-on-Tees on 27 September 1825 to watch the arrival of a steam train, it witnessed history in the making. The slow, 25-mile journey from the collieries near Shildon marked not just the opening of the Stockton & Darlington Railway but also the dawn of the railway age. Horse-drawn waggonways were already a familiar sight in the industrialising north-east of England, so the concept of a railway was not entirely new. This, however, was an acknowledged landmark in the evolution of public transport: the world's first permanent, locomotive-powered, passenger-carrying railway line. Few, if any, of those first spectators could have predicted how far the railway network would develop by the end of the century (see page 103), or how important it would become.

The initial proposals to create a railway line in the area had encountered strong opposition from powerful aristocratic landowners, who resisted its incursion onto their estates. An Act of Parliament authorising the Stockton & Darlington Railway Company's chosen route was finally passed in 1821. That same year, the company appointed the steam engine pioneer, George Stephenson, as its chief engineer. He decided to upgrade the railway's design so that locomotive-driven trains, as well as horse-drawn waggons, could run along it. This map, printed and published in 1822, reflects his resurvey of the route: the line authorised in 1821 is marked in blue, with Stephenson's alterations in red. The sections below the map, showing the height of the line, became a common feature of 19th-century railway maps. Parliament approved both the revised route and the use of steam power in 1823.

Coal was the driving force behind the railway both literally, as the fuel used to stoke the steam engines, and as the chief reason for its construction. The route connected the coalfields near Bishop Auckland (the area tinted pale blue on the map) with the manufacturing centres of Darlington and Stockton, where the coal powered the towns' burgeoning industries. It was freight services transporting this 'black gold', not passenger trains, which made up most of the traffic during the new railway's first few years.

Operations on the line during those early days remained very different from today's railways, lacking modern timetables, signalling or even proper station platforms. Regular passenger services were initially formed of horse-driven coaches, although steam locomotives were used for hauling freight. Not until 1833 did the route become steam-only. This has led some people to question the status of the Stockton & Darlington Railway as the true birth of rail travel. The Liverpool & Manchester Railway, which ran regular steam-driven passenger services from its inception in 1830, was arguably the first genuine public railway in the modern sense. Regardless of how it began, the revolutionary impact of this mode of transport on human communications and mobility can scarcely be overestimated. The coming of railways transformed landscapes and lives not just in the north-east of England but throughout the world.

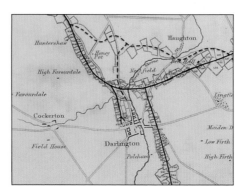

RAILWAY TOWN: The market town of Darlington, lying roughly halfway along the original route, is closely associated with trains. For many years it was home to three railway manufacturing firms and it remains a railway junction. The yellow line shown here is a proposed branch to the Yorkshire town of Croft-on-Tees.

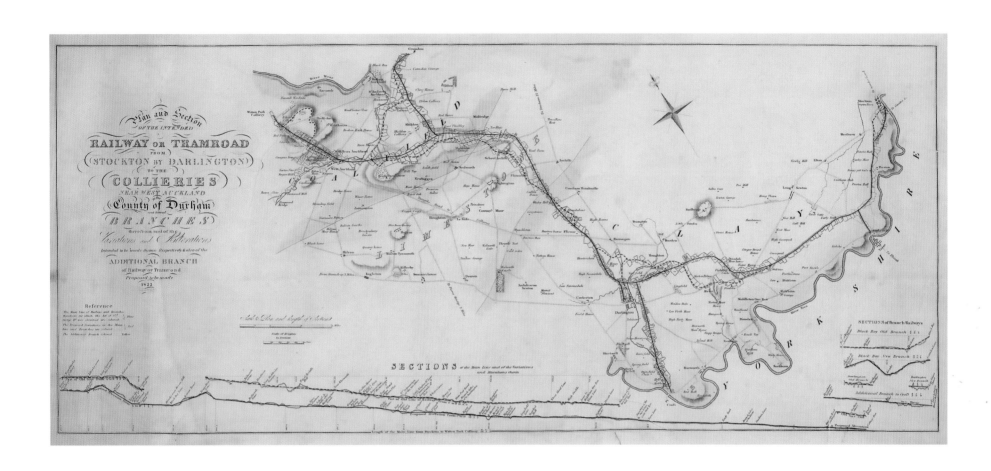

HISTORIC JOURNEY: An artist's impression of the first passenger journey on the Stockton & Darlington Railway.

OPENING OF THE FIRST ENGLISH RAIL-WAY BETWEEN STOCKTON AND DARLINGTON, SEPT. 27TH, 1825.

# At a curious angle

LAKE OF THE WOODS, UNITED STATES OF AMERICA AND CANADA, 1872–1876

The international boundary separating Canada and the United States of America is the world's longest border shared between a pair of countries. The fact that much of the boundary follows the line of latitude 49°N has given rise to the use of 'the 49th parallel' as a metaphor when discussing differences between the two nations. The status of this arbitrary line dates from the London Convention of 1818, which established it as the boundary from the Lake of the Woods in the east to the Rocky Mountains in the west. This portion of the border remained undefined on the ground until the British and Americans carried out a joint survey to delimit it between 1872 and 1876.

Drawing a boundary line on a map is relatively easy, even if persuading people to agree to it is not. Transforming that line into a physical presence is an altogether different challenge. This particular boundary is not in fact quite as straight as it was intended to be, or as it appears on this map. The survey party cleared a strip of land 20 feet wide and built hundreds of boundary stones along it to mark the border. The tools available to the men engaged in the hard physical labour of cutting a swathe through the forest and setting the stones in place did not allow for precise measurement. Thus, rather than following the parallel exactly, the physical border meanders gently back and forth on its 800-mile path from east to west.

This map, which was made for the Boundary Commissioners in about 1874, illustrates a rather more significant anomaly in the border. The map covers the section of the boundary east of the Red River. On its right-hand side is a wedge-shaped area on the western shore of the Lake of the Woods known as the Northwest Angle. Despite being situated on the Canadian side of the lake, the Angle forms a practical exclave of Minnesota in the United States.

This geographical quirk has its origin in an historical misunderstanding. Under the terms of the 1783 Treaty of Paris – which was signed before the area had been mapped in detail – the international boundary was supposed to run due west from the north-westernmost point of the Lake of the Woods until it reached the Mississippi River. This subsequently proved to be impossible: a boundary running westward from this point would never reach the Mississippi because its source lies some distance south of the lake. Accordingly, it was resolved that the boundary should instead run due south from the lake's north-westernmost point as far as the 49th parallel, before turning west. The lake also proved to be less regular in shape than European explorers had thought initially. When accurate measurements were made, the agreed boundary was discovered to take the curious form that we see today.

CUTTING THROUGH THE FOREST: This photograph showing the clearance of trees from the boundary line at the Northwest Angle also comes from the Boundary Commission's records.

DISPUTED TERRITORY: Land on the Canadian side of the border was hotly disputed between the provinces of Manitoba and Ontario. An Act of Parliament passed in London in 1889 placed the provincial boundary significantly further east than the line depicted here.

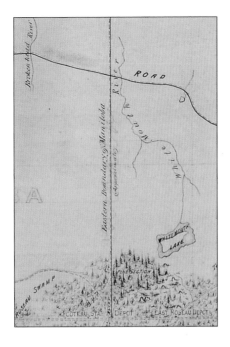

WOODS OF THE LAKE: These delicately sketched tree symbols represent the region's thickly forested terrain, which inspired the name Lake of the Woods.

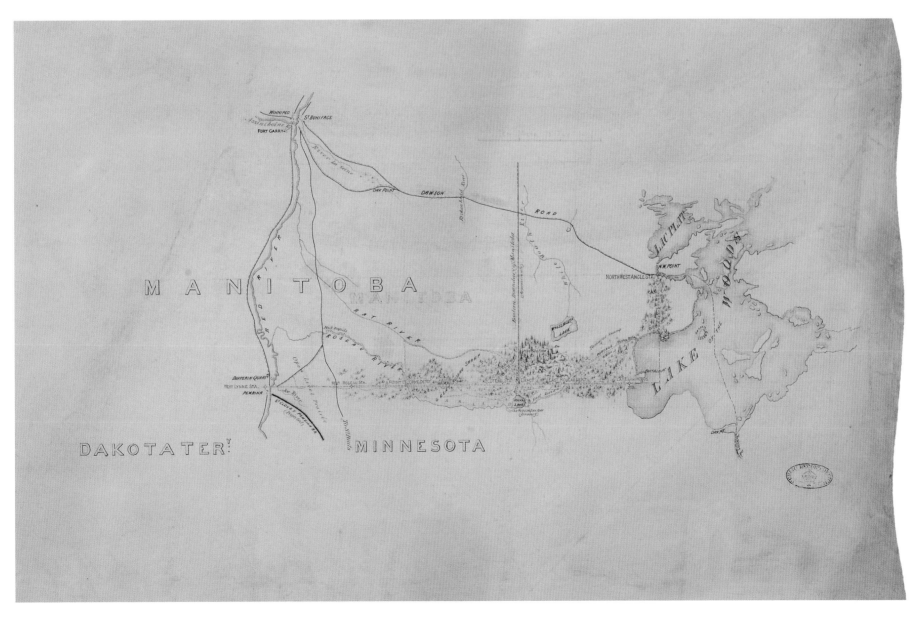

# New Territories

HONG KONG, 1898

When China lost the First Opium War (1839–1842) against the British, her defeat marked the beginning of a 'century of humiliation' during which she was dominated politically and economically by western countries and Japan. Under a series of unequal treaties, various nations received trading privileges and land grants within China, generally in the form of foreign concessions and leased territories. Among the requirements of the 1842 Treaty of Nanjing, which ended the Opium War, was the cession of Hong Kong Island to the United Kingdom. The colony grew into a flourishing free port and centre of international trade. Its population increased rapidly and in 1860 it was extended to include the neighbouring Kowloon peninsula.

Towards the close of the 19th century, Hong Kong was becoming overcrowded and its authorities wanted a further extension. The United Kingdom's government also had political reasons for desiring its expansion: a larger Hong Kong would make the colony easier to defend and safeguard British influence in China at a time when French, German and Russian interests in the region were strengthening. On 9 June 1898 British and Chinese representatives signed a treaty offering the United Kingdom a 99-year lease of an area subsequently termed the New Territories. This area (shown in white on the map) was more than four times the size of the existing colony (coloured pink) and comprised some 200 islands, as well as a slice of the mainland. The lease began almost immediately, on 1 July, but formal ratifications of the treaty were not exchanged until 8 August and colonial officials did not take effective control over the New Territories until the following spring.

Each government signing a treaty customarily retains an official copy for its archives. Our copy of the 1898 treaty consists of separate handwritten texts in Chinese and English, each signed and sealed by both parties. The two texts are bound together with this bilingual map, which is drawn neatly on tracing linen. Oddly for such a formal document, the scale of 1 inch to 4.96 miles is noted lightly in pencil. The treaty specifies that the colony's new boundary (shown as a dashed line) had been agreed in principle, but a survey on the ground was required to fix it precisely. Other notable provisions include permission for Chinese warships to sail in the leased waters of the Deep Bay and Mirs Bay, and measures to minimise disruption to local officials and the Chinese population affected by the change in sovereignty.

British officials in 1898 apparently thought that 99 years was tantamount to forever and regarded the lease as effectively permanent. History was to prove them wrong. In 1984 the British and Chinese governments signed a joint agreement that the United Kingdom would restore not just the New Territories but the whole of Hong Kong to China when the lease expired. Accordingly, Hong Kong's status as a British Crown colony came to an end at midnight on 1 July 1997, exactly 99 years after the lease had officially begun.

STAMP OF APPROVAL: This seal was applied to the treaty by the Chinese delegation.

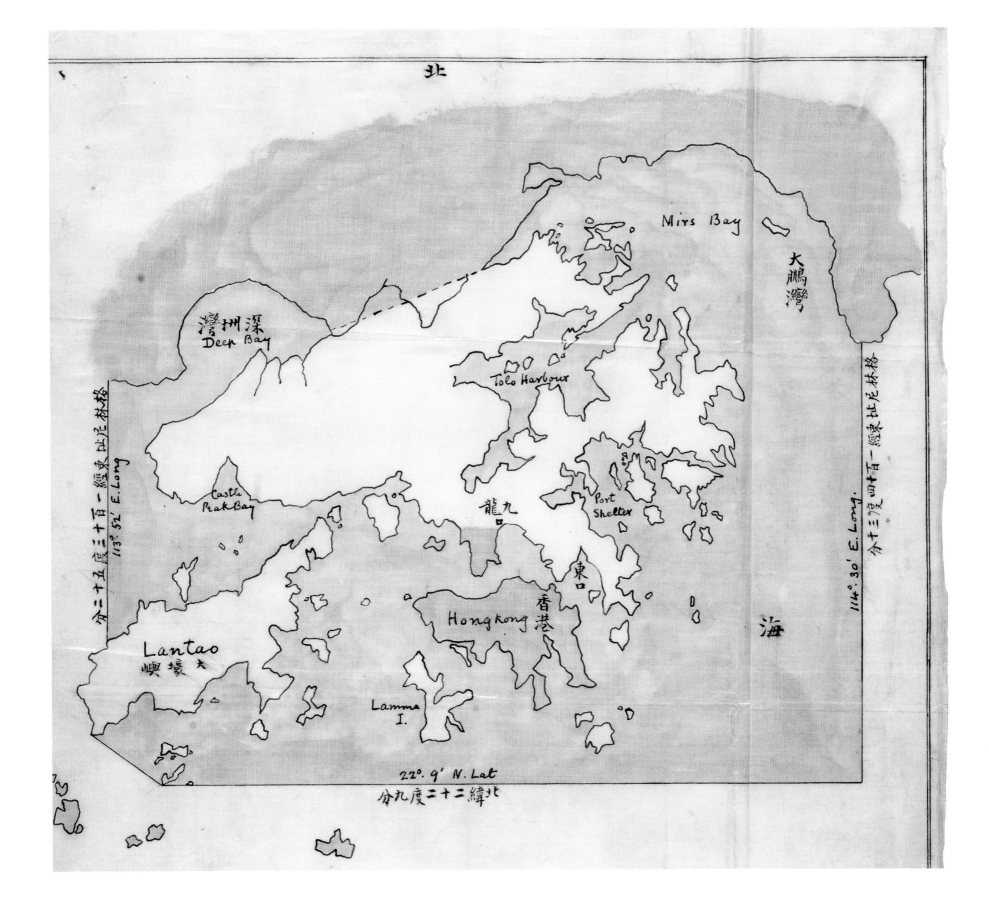

北

Mirs Bay

大鵬灣

深圳灣
Deep Bay

Tolo Harbour

113° 52′ E. Long 粉嶺東正一百三十五度三十分

114°. 30′ E. Long. 粉嶺東一百十四度三十三分

Castle Peak Bay

龍九

Port Shelter

Hongkong 香港

東口

海

Lantao
大嶼山

Lamma I.

22°. 9′ N. Lat
北緯二十二度九分

# An evil empire? THE BRITISH EMPIRE, 1915

First impressions are not always reliable. Superficially, this map appears to represent the world as a British schoolchild of the early twentieth century would have been taught to perceive it. The conventional, cylindrical projection creates a rectangular shape, with the Americas to the viewer's left, Asia to the right, Africa occupying the middle and the United Kingdom in prime position above the centre. The title proclaims the map's subject to be the British Empire, and the mother country and her colonies are tinted the traditional pink.

This was not, however, a map intended to glorify Britain's imperial power but exactly the opposite. The first clue to this is that it was published in Berlin in 1915, when the British and German Empires were engaged on opposing sides in the First World War. Its publisher, Dietrich Reimer, was an established company with specialisms in geography, archaeology and art, whose authoritative reputation would have aided the effectiveness of this item as a piece of anti-British propaganda.

The purpose of the map is revealed by a closer examination of its legend, which purports to list how the United Kingdom acquired each part of her empire. Most of the descriptions are best considered as partial truths: whilst falling short of outright lies, they lack any attempt at balance, completeness or even strict accuracy. For example, Canada is noted as 'captured from France', when only parts of that dominion were ever settled as French colonies. Writing in the December 1917 issue of the *Geographical Journal*, a British commentator observed sarcastically that the term 'legend' was an apt label for such dubious statements.

It was this same article that led the Colonial Office to acquire this map. An official who had read the journal asked the Stationery Office to obtain a copy of it because he realised that German propaganda relating to British possessions overseas would be valuable to his department. In May 1918, the map was duly delivered to the Colonial Office, where it was filed with a typed copy of part of the article glued to its back.

A notable omission from the map as printed is any indication of the colonies held by other European powers. These blank spaces of non-British territory cleverly convey the false impression that land-grabbing imperialism was a uniquely British phenomenon. In fact, Germany had built up a significant overseas empire of her own, chiefly in Africa and the Pacific Ocean, most of which would be captured by her opponents as the war progressed. After Germany's defeat, the administration of her former colonies – and of some territories lost by Turkey, her ally – was divided among other countries, including the United Kingdom. Some, though not all, of these post-war additions to the British Empire have been marked on this map in pencil. We suspect that Colonial Office officials found this subtle subversion of the map's original purpose rather satisfying. It is certainly a neat reflection of how the balance of global power had shifted in Britain's favour.

THE MISSING LINK? Most of the former German East Africa came into British hands after the war. Administered under the name Tanganyika, this territory became the last link within a continuous stretch of British-controlled land spanning Africa from Cairo to Cape Town. It now forms the greater part of Tanzania.

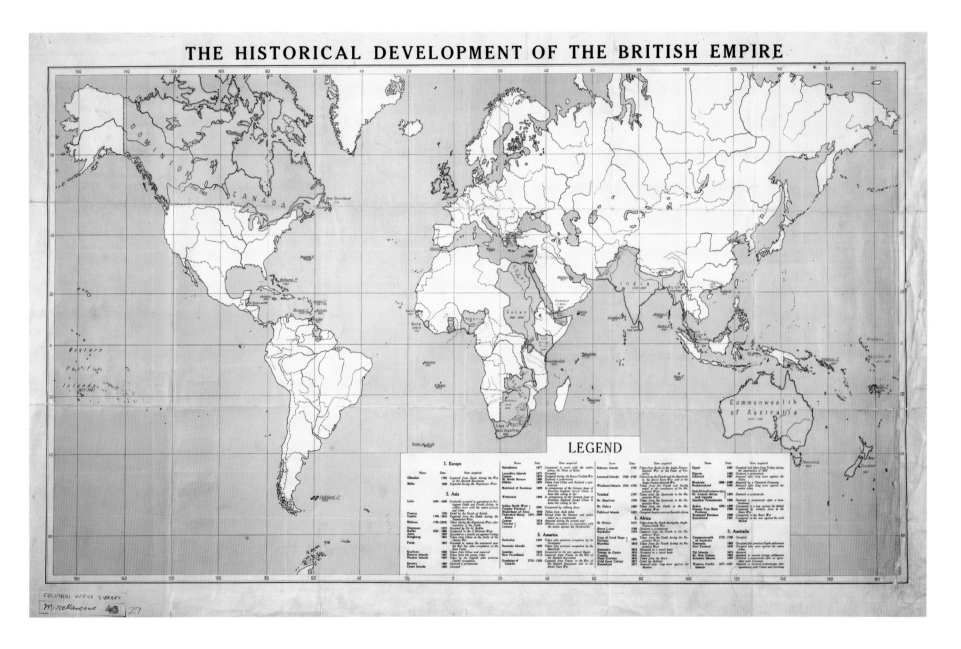

# THE HISTORICAL DEVELOPMENT OF THE BRITISH EMPIRE

## LEGEND

### 3. America

| Barbados | 1605 | Taken after previous occupation by the Portuguese |
| Bermuda Islands | 1609 | Taken after previous occupation by the Spaniards |
| Jamaica | 1655 | Conquered in the war against Spain |
| New Foundland | 1713 | Captured from France in the War of the Spanish Succession |
| Dominion of Canada | 1713—1763 | Captured from France in the War of the Spanish Succession and in the Seven Years War |

A FAIR REFLECTION? The legend expresses a distinctly anti-British perspective on three centuries of world history.

# The land of the dead

The people of ancient Egypt hoped that after death they would be granted a second life in the *duat* (or land of the dead). For those who could afford them, elaborate rituals developed around the mummification and burial of corpses. The pharaohs who ruled Egypt spared no expense on the construction and decoration of tombs for themselves and their families, and they were buried with many valuable objects to support their future existence in the duat. In today's world, the ancient Egyptians enjoy a different kind of afterlife: their long-lasting civilisation continues to fascinate many people, from young children to academic researchers.

Although Western society's preoccupation with ancient Egypt dates back centuries, the sensational discovery of the tomb of the pharaoh Tutankhamun in 1922 led to a renewed wave of interest. All of the royal gravesites discovered previously had been robbed of their treasures within a few hundred years of the burial, but Tutankhamun's tomb was virtually intact. It took eight years for a team led by the archaeologist Howard Carter to excavate the burial place thoroughly and remove all of its precious objects. The deaths of several people connected with the excavation (including the Earl of Carnarvon, who had commissioned it) prompted rumours that the tomb had been cursed to deter grave robbers. Such speculation seems to have fuelled, rather than detracted from, 'Tutmania' – a craze for incorporating Egyptian elements into architecture, decorative art and fashion that strongly influenced the evolution of Art Deco style.

Tutankhamun reigned between about 1332 and 1323 BC, during the New Kingdom period which marked the zenith of Egyptian wealth and power. The favoured method of royal burial at this time was to construct hidden tombs by cutting deep into the rocks. The royal necropolis was laid out on the western side of the River Nile, opposite the city of Waset, or Thebes (on the site of modern Luxor), which was the pharaohs' capital for most of this era. This map shows the Valley of the Kings, which includes the burial places of Tutankhamun (number 62), Ramesses the Great (7), and the female pharaoh Hatshepsut (20). The brown contour lines, set two metres apart, reveal the steepness of the landscape.

The map is one of a set of twenty sheets published by the country's official mapmaking organisation, the Survey of Egypt, in co-operation with the Antiquities Department. It was produced in 1926, while the excavation of Tutankhamun's tomb was still underway. Although a formal British protectorate over Egypt had nominally ended four years earlier, the United Kingdom still exerted a strong influence over the Egyptian government, and the Survey was still staffed by a mixture of British and Egyptian personnel. Our copy of the map comes from the Colonial Office's map collection. Its presence there reflects the prominence of archaeology and antiquities in Egypt's relations with the United Kingdom at this time. Now that the British government's original need for the map is over, it has entered its own afterlife, preserved here in the archives as an historical record.

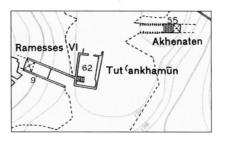

TOMB OF TREASURES: Tutankhamun succeeded to the throne as a child and died in his late teens. His reign was notable for revival of traditional Egyptian religion, which had been altered radically under his predecessor Akhenaten.

# THE THEBAN NECROPOLIS

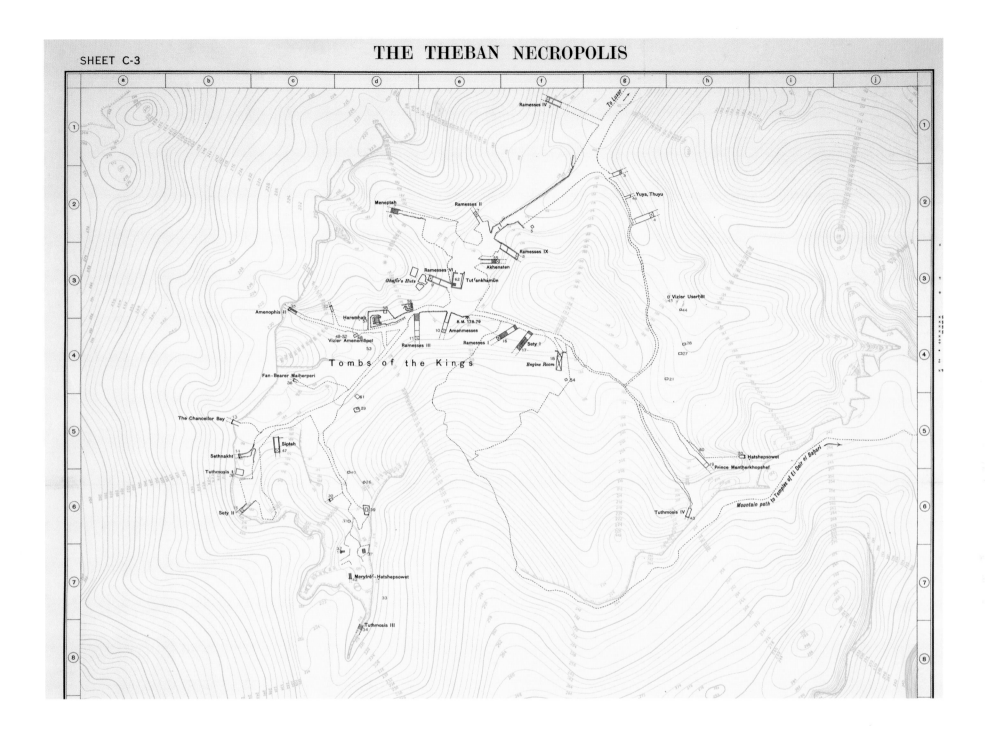

Ramesses IV

To Luxor

Yuya, Thuyu

Meneptah

Ramesses II

Ramesses IX

Akhenaten

Ramesses VI

Ghafir's Huts

Tut'ankhamūn

□ Vizier Userhēt

Amenophis II

Haremhab

B.M. 178-79

Vizier Amenemōpet

Amenmesses

Ramesses III

Ramesses I

Sety I

Tombs of the Kings

Engine Room

Fan-Bearer Maiherperi

The Chancellor Bay

Hatshepsowet

Sethnakht

Siptah

Prince Mentherkhopshef

Tuthmosis I

Sety II

Tuthmosis IV

Mountain path to Temple of El Deir el Bahari

Merytrēʿ-Hatshepsowet

Tuthmosis III

**EGYPTOLOGISTS AT WORK:**

An artist's impression of an
archaeological site.

# 'Peace for our time' CZECHOSLOVAKIA, 1938

Czechoslovakia was one of the new countries created in 1918 when the Austro-Hungarian Empire was dismantled after the First World War. As the old empire had been notably cosmopolitan, encompassing many ethnically and linguistically mixed regions, all of the successor states included substantial minorities within their borders. For instance, many towns and villages in Czechoslovakia, especially in the western part of the country near to its borders with Germany and Austria, were inhabited by ethnic Germans.

In the late 1930s the German government demanded a change to the boundaries of Czechoslovakia that would transfer its majority-German-speaking areas, known as the Sudetenland, to Germany. In late September 1938, Adolf Hitler, the German Chancellor, met the leaders of the United Kingdom, France and Italy in the German city of Munich to discuss this territorial claim. During the discussion, Hitler gave this map of Czechoslovakia, illustrating his demands, to the British Prime Minister, Neville Chamberlain. The map is colour-printed to show the distribution of Czechs (beige), Slovaks (pink), Hungarians (yellow), Poles (mauve), Germans (blue) and Ukrainians (orange) throughout the country. The four areas that Hitler intended to annex are marked with hand-drawn dark blue lines and numbered in red in Roman numerals. The other handwritten figures, also in red, indicate the specific dates in early October when the annexations of each part of the Sudetenland were planned to take place.

In the hope of averting war in Europe, Chamberlain and his French counterpart, Édouard Daladier, agreed to Hitler's demands, against the wishes of the Czechoslovakian government. In return, Hitler promised not to annex any territory beyond his existing claims. On Chamberlain's return to London on 30 September, he made a speech describing the agreement at Munich as 'peace for our time' but subsequent events proved his optimism unfounded. The remainder of Czechoslovakia disintegrated over the coming months as a number of areas came under Hungarian, Polish and German control. Less than a year after the Munich agreement, Hitler broke his promise not to claim additional territory. On 1 September 1939, German troops invaded western Poland, sparking the Second World War.

When the war ended in 1945, Czechoslovakia was re-established. Its pre-Munich boundaries were restored in the west, although it lost some territory in the east. In a 'velvet divorce' between the country's eastern and western halves at the beginning of 1993, Slovakia and the Czech Republic became two fully independent states. By a quirk of history this new international boundary was also an old one: it can be traced back to the boundary between the medieval kingdoms of Hungary and Bohemia.

As for this map, Sir Horace Wilson, a senior civil servant who had accompanied Chamberlain to Munich, suggested that it should be given to the British Foreign Office for safekeeping. It was added to the Foreign Office's map collection. We now keep it in a special, high-security storage area, reserved for The National Archives' most important and valuable records.

SIR HORACE'S SUGGESTION: This letter, which is attached to the map, explains how it came into the custody of the Foreign Office. Winston Churchill.

NEVILLE CHAMBERLAIN

PORTRAIT OF A PRIME MINISTER: This caricature of Neville Chamberlain comes from the records of the Ministry of Information. Chamberlain resigned as Prime Minister in May 1940 and was replaced by Winston Churchill.

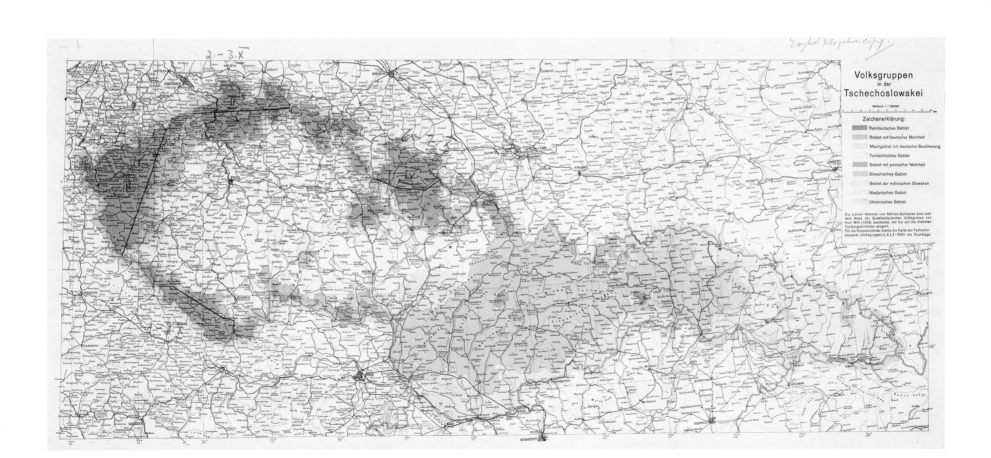

# Reports of 'gas chambers in a wood'

OŚWIĘCIM (AUSCHWITZ), POLAND, C.1942

The Nazi German regime followed a doctrine of racial purity which characterised the 'Aryan' Teutonic and Nordic people as superior to others. The regime targeted many groups – including Roma (gypsies), Slavs, homosexual men, people with physical or mental disabilities, and those who opposed Nazism for political or religious reasons. Jewish people, however, were singled out for particular hatred, discrimination and eventually ethnic cleansing. Up to six million people – one third of the world's Jewish population – were killed in the Nazis' death camps, an event that has become known as the Holocaust.

The largest and most notorious location of this genocide was a complex of three main concentration camps and several sub-camps in the vicinity of Oświęcim (known in German as Auschwitz). The town lay in the region of Upper Silesia which had been annexed to Germany following her 1939 invasion of Poland. This photostatic copy of a sketch map shows the area as it was in early 1942, when much of the camps' infrastructure was still being developed. Point C – described in the accompanying papers as 'gas chambers in a wood' – is identifiable as the site of Auschwitz II (Birkenau), where mass killings of inmates took place from 1942 onwards. The sketch was based on a description supplied by an anonymous prisoner held in another part of the complex, perhaps one of the blocks on the right-hand side. Apart from a small inset in the top left corner, which shows the accurate relative positions of the main locations, the map is not drawn to scale.

By the latter stages of the Second World War, intelligence about the Holocaust had begun to reach Germany's opponents. In the summer of 1944 the Jewish Agency for Palestine proposed a scheme to bomb the gas chambers at Auschwitz and other camps, and the neighbouring railway lines. The intention was to prevent more Jews from being transported to their deaths. Although senior British politicians supported this plan in principle, they rejected it for practical reasons. The region was too far from any Allied airfields for the raids to be carried out safely. There was also thought to be too little information available about the topography of the region and the layout of the camps for bombs to be targeted accurately; there was a risk that an air raid would kill the inmates rather than saving them. Doubts were also raised as to whether any Allied action could deter the Nazis from their ideologically-motivated slaughter.

To support its case for the bombing scheme, the Jewish Agency had obtained this map of Auschwitz and a rough plan of the camp at Treblinka from the London-based Polish government-in-exile and forwarded them to the British Foreign Office. In late September, it was discovered that the two maps had not been passed to the Air Ministry as intended, and hence had not been taken into account when evaluating the proposal. Officials at the Foreign Office decided that the maps were insufficiently detailed or recent enough to have influenced the outcome. Historians continue to disagree about whether the British government's decision to reject the bombing plan was correct, and about the accuracy and completeness of the intelligence that informed it.

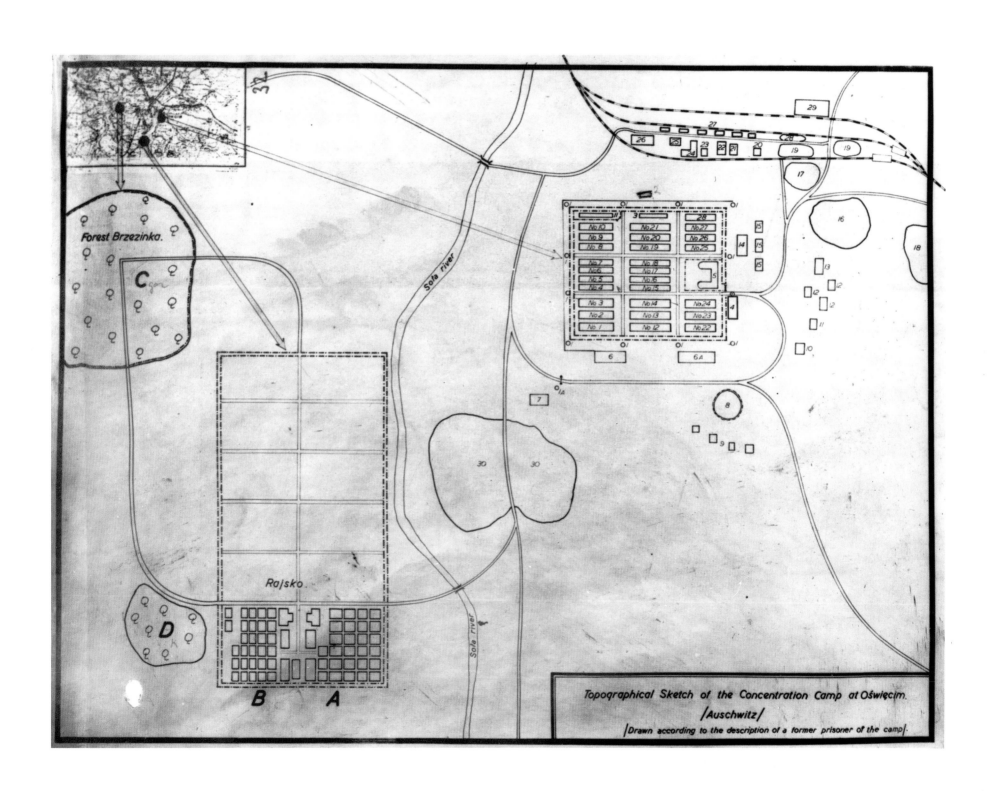

Forest Brzezinka.

C

D

Rajsko.

B  A

Sola river

Sola river

30  30

No 10  No 21  No 27
No 9  No 20  No 26
No 8  No 19  No 25

No 7  No 8  5
No 6  No 17
No 5  No 16
No 4  No 15

No 3  No 14  No 24
No 2  No 13  No 23
No 1  No 12  No 22

6  6A

7  1A

8

29
27
26  23  28  20
24  22  21  19  19
17
16  18
15
14  15
15
13
12  12
12
11
10

9

Topographical Sketch of the Concentration Camp at Oświęcim.

/Auschwitz/

/Drawn according to the description of a former prisoner of the camp/.

# The last place on Earth  ANTARCTICA, 1953-1957

Many old maps of the world show a vast, hypothetical southern continent labelled Terra Australis Incognita, a Latin name meaning 'unknown southern land'. By the end of the 18th century, European geographers realised that speculation about this land had not been accurate: the voyages of Abel Tasman, James Cook and others had demonstrated that Australia and the islands of the south-western Pacific Ocean did not form part of any such supercontinent. Indeed, the name of Australia – formally adopted in 1824 – reflects the early 19th-century belief that it was the world's southernmost continent. By this time sightings of the landmass that we now know as Antarctica had already begun, although it was not until 1840 that the American naval officer Charles Wilkes proposed the existence of 'an Antarctic continent'. The first use of the name Antarctica in print is credited to the Scottish mapmaker John Bartholomew in the 1890s.

Serious exploration of the continent's interior began in 1897 and continued for the next 25 years. A Norwegian expedition led by Roald Amundsen reached the South Pole in December 1911, beating a rival British party by a few weeks. By that time the British had already laid claim to a slice of Antarctica, and several other countries followed suit over the following decades. When this map was produced in 1953, seven countries – the United Kingdom, New Zealand, France, Norway, Australia, Chile and Argentina – maintained land claims, with only Marie Byrd Land left unclaimed.

Although the map primarily shows the political situation on land and was not intended for navigation at sea, it is technically a chart, prepared by the Hydrographic Department of the British Admiralty. The annotations marking British dependencies in the South Atlantic Ocean were probably added in March 1957. The map illustrates a Commonwealth Relations Office file relating to British policy in Antarctica in the late 1950s, when the continent's political importance reached its peak. The file reveals that the United Kingdom had considered withdrawing her claim to part of the continent but decided not to do so, fearing a negative impact on her strategic interests in the area and on the claims of her key allies (and former colonies) Australia and New Zealand. Against the backdrop of the Cold War, it was also thought that additional nations – including both the United States and the Soviet Union – were likely to stake their own claims to territory.

The Antarctic Treaty of 1959 averted the potential for conflict. It permitted no additional land claims but neither approved nor denied the legitimacy of the existing ones. Thus, the seven territorial claims extant in the 1950s remain the same today. The treaty also barred military installations, weapons testing and nuclear waste from the continent, reserving its use for peaceful, scientific purposes. Antarctica remains the only 'unspoilt' continent with no permanent human population, although a few thousand people live there temporarily in research stations. International co-operation has kept Antarctica conflict-free for more than six decades, a truly historic occurrence.

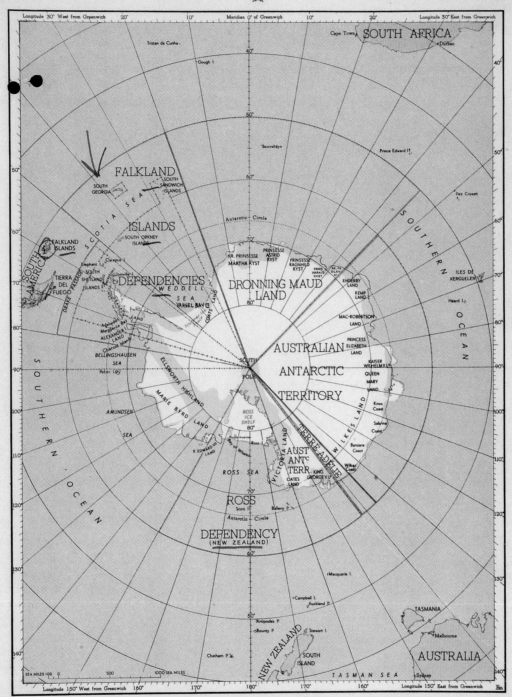

# THE ANTARCTIC
## GENERAL MAP SHOWING TERRITORIAL CLAIMS, 1953
### KEY TO SYMBOLS USED

▭ AREAS CLAIMED TO HAVE BEEN SEEN BY EXPLORERS

BRITISH TERRITORY ——————        ARGENTINE CLAIMS ——————
NORWEGIAN TERRITORY ——————        CHILEAN CLAIMS ——————
FRENCH TERRITORY ——————        (Northern limit undefined)
ICE FRONT ~~~~~~

AZIMUTHAL EQUIDISTANT PROJECTION

Prepared by the Hydrographic Dept of the Admiralty, 18th April 1946 under the Superintendence of Rear-Admiral A. G. N. Wyatt, Hydrographer.

New Editions 25th September 1953

D.6479

# CHAPTER EIGHT

# WORLDS OF
# imagination

Government is a serious business but this does not mean that its records are dull and uninteresting. On the contrary, the historical records of the British government preserved at The National Archives are full of life and human interest. Throughout this book we have featured maps that embody and reflect the stories lying behind government recordkeeping and the events of the past. Although government bodies have made, used and acquired maps for practical – and sometimes deadly serious – reasons, this does not prevent many of our maps from displaying the attractive, quirky or even humorous aspects of cartography.

In fact, all maps are exercises in human creativity. Since no map can record every detail of the places that it depicts, cartographers must make creative choices, either consciously or unconsciously selecting what to include and what to exclude. In this sense, every map combines a world of observation with a world of imagination. The balance between the two varies from mapmaker to mapmaker and from map to map. Whilst observation, measurement and accuracy are often paramount for official purposes, in other contexts, cartographers can allow their imaginations to run more freely.

Perhaps the most obvious way that a map can reflect its maker's sense of creativity is by combining practical information with an aesthetically-pleasing appearance. Some maps include distinct decorative elements, such as elaborate borders or fearsome sea monsters; others are simply examples of attractive and well-balanced design. Although cartography is no exception to the rule that beauty is in the eye of the beholder, many of the maps included in this book are genuinely beautiful or visually striking.

Sometimes the exercise of creativity results in a map with an unusual format or physical makeup. Usable objects incorporating maps into their designs – sometimes called cartofacts or cartifacts (a portmanteau of cartographic and artefacts) – are

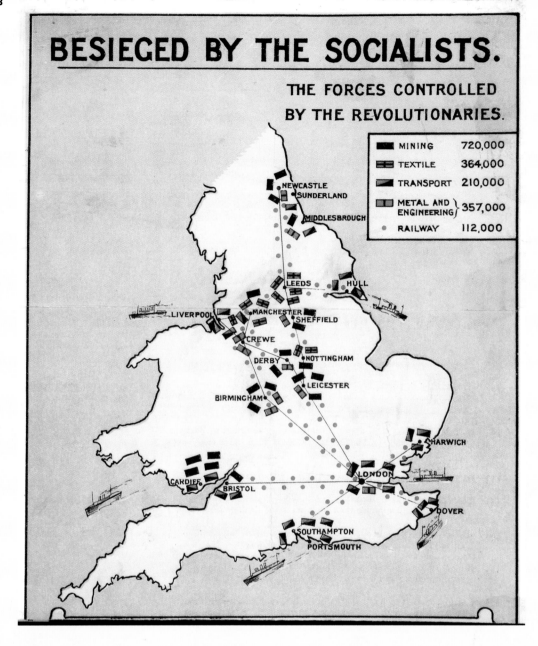

**3**

# BESIEGED BY THE SOCIALISTS.

## THE FORCES CONTROLLED BY THE REVOLUTIONARIES.

| | |
|---|---|
| MINING | 720,000 |
| TEXTILE | 364,000 |
| TRANSPORT | 210,000 |
| METAL AND ENGINEERING | 357,000 |
| RAILWAY | 112,000 |

**1** THE MAP AS ART: This sculpture, entitled 'There be monsters' sits in the grounds of The National Archives at Kew. Its design was inspired by some of our historical maps.

**2** A DECORATIVE DETAIL: This sketch of a penguin shooting arrows at a sea monster illustrates a track chart of a Second World War naval battle in the Barents Sea, north of Norway. The penguin has evidently strayed a long way from its natural home in the southern hemisphere.

**3** A POINT OF VIEW: Issued by the Anti-Socialist Union of Great Britain in 1912, this piece of propaganda is a clear example of the map as a political statement.

the logical extreme of this idea. In most of these cases, the function of the map element is aesthetic rather than practical. Instead of incorporating decoration, the map itself forms the decoration.

Other maps are imaginative in the very different sense of depicting places that are not real. These range from maps of wholly imaginary places to maps showing planned or proposed changes to real places that, for one reason or another, have never become reality. Somewhere in between these two extremes lie maps of real places portrayed inaccurately. As cartographers in past centuries often had much more limited access to geographical information or knowledge than we do today, it is not surprising that mistakes crept in. Some mapmakers used their imaginations to fill what would otherwise have been blank spaces on their creations.

A more subtle aspect of imagination in cartography is the creative purpose or message inherent in a great many maps. Far from being made for general purposes, like the published products of the Ordnance Survey and other ordinary topographical mapping, many maps have been designed and made with the needs of advertising, propaganda, satire or simply entertainment in mind.

Various different facets of imagination and creativity may, of course, be combined. For instance, a satirical map may also depict an imaginary place, as in the map of 'Green Bag Land' on page 227 or an advertisement may incorporate a map into a bold and clever design, such as the poster on page 239 In this final chapter, we explore a selection of maps that reflects these diverse aspects of the cartographic imagination.

**4**

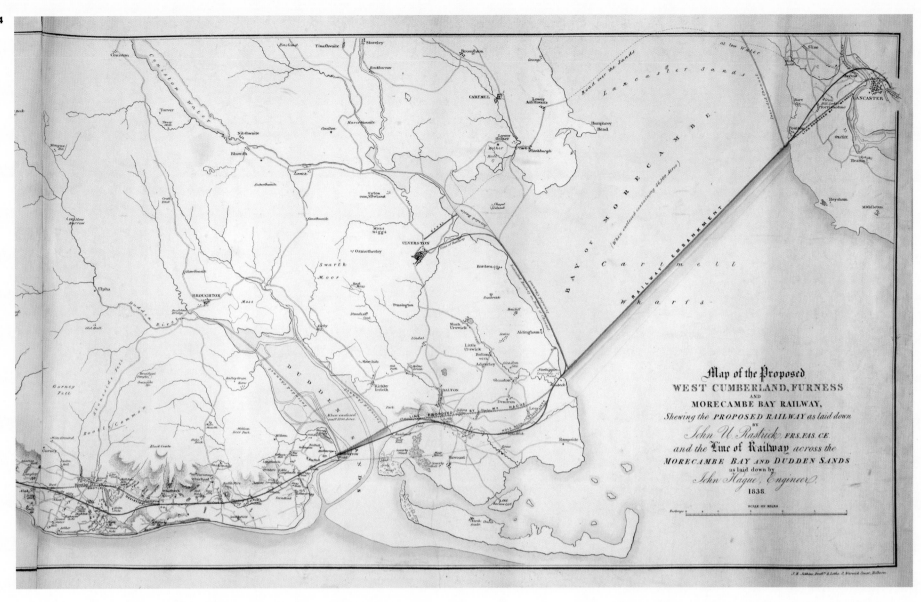

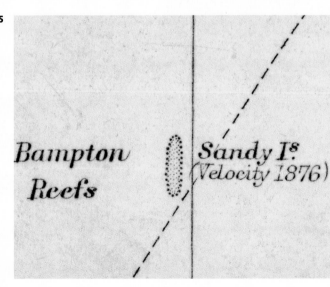

**4** AN UNSUCCESSFUL PROPOSAL: The possibility of building a railway embankment across Morecambe Bay, in the north-west of England was considered in the 1830s. No such railway line has ever been built.

**5** A PHANTOM ISLAND: Sandy Island was supposedly sighted near New Caledonia by Captain Cook in 1774. It subsequently appeared on many maps, including this British Admiralty chart of the south-west Pacific Ocean, dating from 1885. The island's non-existence, and its removal from Google Maps, featured in news headlines around the world in November 2012.

# Brave new world    THE AMERICAS, c.1700

In the past, accurate information about faraway places was often difficult or impossible to obtain. One result of this is that old European maps of other continents often presented geographical conjecture as fact. Perhaps the most notorious of these errors was the island of California. Although it had been correctly depicted as part of the American mainland on 16th-century printed maps, a misconception arose early in the 17th century that California was an island and, for more than a hundred years, many European mapmakers drew it that way.

As geographical knowledge increased during the early 18th century, debate raged about whether California was truly an island. Despite a reliable eyewitness report from a Spanish missionary, Eusebio Kino, that Lower California was a peninsula, many leading cartographers, such as Herman Moll, continued to prefer the traditional island shape. The Dutch publisher Pieter van der Aa offered his customers a choice between maps portraying California as an island and maps depicting it as a peninsula. In 1747 King Ferdinand VI of Spain settled the question by issuing a royal decree stating that California was not an island. The geographical myth of a detached California soon became a footnote of history.

Around the turn of the 18th century, when this map of North and South America was engraved, the 'New World' was still a distant and largely unknown place for many Europeans. The island of California is just one of many speculative pieces of information featured here. Many details, from the courses of rivers to the positions of Pacific Islands, are highly inaccurate and several of the boundaries have little, if any, basis in fact. The fictitious realm of *Terra Esonis* (corresponding roughly to Alaska and eastern Siberia) occupies a large, almost blank expanse in the northwest. To modern eyes, such imaginative details enhance the impression of the map as an old and beautiful object. Like its style of engraving, hand-coloured outlines and text in Latin, they are part of its charm.

The map was created by the Dutch cartographer Carel Allard, who headed his family's business in Amsterdam between 1691 and 1706. Our example forms part of a two-volume atlas compiled from the work of several Dutch mapmakers, chiefly Frederick de Wit, and sold in London by Christopher Browne. The atlas formerly belonged to the Board of Customs, which was responsible for controlling imports and exports and preventing smuggling. The fact that it was owned by the government and, we believe, had a practical function as a work of reference, reminds us that this map is more than just an historical curiosity. Although its claim to be the 'very latest map' (*recentissima tabula*) of the Americas is hyperbole, maps of this kind were actually the best – and often only – means of access to geographical information about distant lands. The New World portrayed here may strike us today as a world of imagination but, for those who first used it, this map was the gateway to a world of knowledge.

ARTIST'S IMPRESSION: This fanciful scene surrounding the title depicts a Native American woman, her male slave or servant and a variety of 'exotic' flora and fauna, including a very odd-looking alligator or oversized lizard. Tucked away in the lower left-hand corner of the map is the partial outline of *Nova Zeelandia* (New Zealand), named by Abel Tasman after the Dutch province of Zeeland.

Recentissima
NOVI ORBIS.
sive
AMERICÆ,
SEPTENTRIONALIS
et
MERIDIONALIS
TABULA
EX OFFICINA
CAROLI ALLARD
Cum Privilegio Ordinum
Hollandiæ et Westfrisiæ

# A royal scandal  'GREEN BAG LAND', 1820

King George IV was not a success as a husband. As Prince of Wales, he first went through a marriage ceremony with Maria Fitzherbert, but this marriage, contracted without permission from his father or the Privy Council, was not legally valid. His second, official marriage, to his cousin Princess Caroline of Brunswick-Wolfenbüttel, was unhappy from the start. By the time George became king in January 1820, they had been married for nearly 25 years but had lived apart for most of that time and had been publicly estranged for almost a decade. The couple's only child, Princess Charlotte, had died aged 21. George was openly unfaithful to Caroline and many people believed that she was also unfaithful to him. The most senior servant in her household, Bartolomeo Pergami, was rumoured to be her lover.

George decided to divorce Caroline, which required an Act of Parliament. The Pains and Penalties Bill, which proposed to end their marriage and prevent Caroline from using the title of queen, was introduced to Parliament on 5 July 1820. Obtaining a divorce depended on proving that Caroline had committed adultery, so the bill effectively placed her on trial, with Parliament serving as both jury and judge. George's plan failed. Public sympathy lay with Caroline and, although the bill passed narrowly in the House of Lords on 6 November, it had little chance of passing in the House of Commons and the government withdrew it. Caroline retained the status of queen accepting an annual allowance of £50,000. Her victory did not last long; she died the following year, aged just 53.

The scandal surrounding the royal couple inspired a vast amount of satire in various forms. Some satirists supported George and others Caroline, with the balance of opinion in the queen's favour. This map of the imaginary 'Green Bag Land' comes from a volume of cartoons collected together for the Treasury Solicitor, a small government department supplying legal advice to other branches of government. The map was published by the London bookseller Joseph Onwhyn but its author is unknown.

Several of the map's features refer to notorious aspects of the trial. The 'green bags' were two sealed bags of evidence against Caroline, the phrase 'non mi ricordo' ('I do not recollect') was frequently used by a prosecution witness under cross-examination, and the boot-shaped 'Country of Lies' is Italy, where Caroline's adultery with Pergami was alleged to have occurred. Some of the more obscure features, including 'the House that Jack Built' and 'the Matrimonial Ladder' allude to satirical poems by William Hone, an outspoken supporter of political reform.

The landscape of Green Bag Land results from a symbiotic relationship between satire and cartography: the map is both inspired by satire and a vehicle for the satirist's art. This parallels the equally symbiotic relationship between the queen and radical politics. Despite having no interest in politics herself, Caroline found radical politicians and their supporters useful allies in her struggle against her husband. In turn, many radicals, however much or little sympathy they had with Caroline personally, found her a useful, temporary figurehead in their campaigns for political reform.

MUD-SLINGING This cartoon captures the acrimonious nature of the royal couple's relationship.

# MAP OF GREEN BAG LAND.

## Description.

*Boundaries.*]—This newly discovered Land is bounded on three sides by the Country of the Bulls, from which it is separated by a *Great Wall*, well planted with Cannon to protect it from incursions, and on the fourth side it is bounded by the Sea.

*Cities.*—The Capital of the Land is the surprising City of *Non mi Ricordo*, which may justly be termed the eighth wonder of the World; it is divided into the upper and lower, and it is here that the Great Fair or Market for sale of *Non mi Ricordos* is carried on.

To the north of the City is an extensive Province, formerly used to grow Cotton, but now appropriated to a *Crop of Green Bags*, which yield an extraordinary Fruit called *Non mi Ricordo*, from whence both the City and the Colonists take their name. Contiguous to this is the Colony of *Double Entendres*, who also drive a fine trade by their commodities, and almost rival the *Non mi Ricordos* by the efficacy of their *Double Entendres*.

*Rivers.*]—A great variety of fine Rivers water this wonderful Country, the principal of which are the following :—the *Waters of Oblivion* take their rise in the Palace of the Great Hum, and after watering the City, and fertilizing the respective Colonial Provinces, empty themselves in the great Washing Tub of the *Non mi Ricordos*. The Waters of this River are so bitter that it would be death to those who drink it, were not its noxious qualities dissipated by the River of Gold, which forms a junction with it a little before entering the City.

The *River of Gold* takes its rise in the Country of the Bulls, in the City of Industry, and flows into the Green Bag Land through one of the *Cannons* in the Great Wall. After forming an agreeable Lake in the Gardens of the Great Hum, it joins the aforesaid River,

*Printed and Published by J. Onwhyn,*

## Description.

and pursuing similar meanderings sweetens its oblivious Waters.

The *River of Truth* takes its rise in the small Palace of the Sultana Hum, which is situated in the Bull Country, and after passing through the various Cities which appear by the Map to be constructed on its Banks, it flows towards the Great Wall where it miraculously ascends a steep Mountain, and by means of a *Steam Engine* is precipitated through a Printing Press over the Great Wall presenting a beautiful cascade in defiance of the Cannons; it immediately forms a great Lake in the Valley of Despair, and rushing from thence in a torrent into the *lower* division of the City, it entirely disperses the before-mentioned River of Gold and Waters of Oblivion.

*Mountains.*]—The most celebrated Mountains are those in the Bull Country. An immense chain surrounds the Land of Green Bags, and are indeed the only security to the Inhabitants of the Bull Country; they are covered with noble forests of waving Goose Quills which give them a beautiful appearance, and the strata of the Mountains are composed of small particles of metallic ore, which, by a particular configuration in the Machines upon their summits keep the whole Land of Green Bags in a state of anxiety, in as much as they are wholly inaccessible to the range of their Cannon, as well as protect the River of Truth which the *Non mi Ricordos* and other Inhabitants of the Green Bag Land have been endeavouring to turn into another Channel.

Mountains are also observed in the Green Bag Land with similar Machines on their summits, but they are wholly artificial, and as they owe their support *solely* to the *River of Gold* and are *far* from the *River of Truth*, their influence over the Bull people is trifling unless conducted through Cannons, or transmitted by Sabres as at the *Field of Peterloo.*

*Catherine Street, Strand.—Price 1s.*

## References.

1 City of *Non mi Ricordo.*
2 Palace of the Great Hum.
3 Palace of the Sultana Hum.
4 Great Washing Tub of the *Non mi Ricordos.*
5 The High Priest's House.
6 City of Peterloo.
7 City of Industry.
8 City of the Matrimonial Ladder.
9 House that Jack Built.
10 City of the Farce of the Green Bag.
11 City of the Man in the Moon.
12 City of the Dainty Dish to set before a King.
13 Vessels with *Non mi Ricordos*, &c. on their Voyage to the Land of Green Bags.
14 *Non mi Ricordos* metamorphised into Gentlemen, previous to landing in the Colony.
15 *Non mi Ricordos* going to Market.
16 *Double Entendres* going to Market.
17 *Non mi Ricordos* and *Double Entendres*, enjoying the Reward of Industry.
18 One of the High Priests watering the Green Bags.
19 The Lake of Despair.
20 Sultana Hum going to the Great City.
21 An Address of the Bulls going up to Sultana Hum.
22 *Precedents* for the Bulls.

# Geography at your fingertips

Most maps, even those that are not beautiful or colourful, are designed to be used and appreciated visually. Tactile maps made for the use of blind people are inevitably an exception because they are intended not to be seen but to be touched.

In 1839, the Glasgow Asylum for the Blind commissioned what is thought to be the first tactile map created in the United Kingdom. The National Archives holds two copies of this map, each made from a sheet of thick, white paper, embossed so that the lines of the map are raised above the rest of the surface. By tracing the surface with his or her fingertips, the reader can 'see' the outline of the British Isles. Although the Braille system of representing letters with patterns of raised dots had been developed in France in the 1820s, it had not yet come into common use in the United Kingdom, so the text is rendered with raised versions of ordinary Roman letters and Arabic numerals.

John Alston, the director of the asylum, sent these two copies of the map to Lord John Russell, the Secretary of State for the Home Department, on 15 August 1839. In drawing the government's attention to the value – and also the difficulty and expense – of making books and other education material accessible to blind people, Alston hoped to secure a small government grant to support the asylum's work. He described the new map as 'very superior' to the tactile maps previously produced at the New England Institute for the Education of the Blind in Boston, Massachusetts, although, as the New England Institute had published an entire atlas of the United States of America a few years earlier, this claim must be treated sceptically. The map has a deliberately simple design, an impression heightened for a sighted person by its complete lack of colour. It shows little more than the outline of the coasts and selected place names, with no sign of the industrial towns, agricultural land, canals, or new railways that formed such prominent features of the early Victorian landscape. The careful marking of latitude and longitude in the margins suggests an attempt at mathematical accuracy.

Although Alston's map is an impressive object for its time, some aspects of it seem rather puzzling from a 21st century perspective. The selection of place names seems arbitrary: for instance, Bristol is marked but Liverpool is not. Although the dashed lines might easily be interpreted as boundaries, only those separating England from Wales and Scotland correspond closely to any actual administrative divisions. Instead, these lines merely group the names of towns and cities with the dots marking their positions. The lack of any clear distinction between individual places and larger areas, unwittingly implying that Scotland, Glasgow and Dumfries are all roughly equal in size and importance, must have left the map's original users with a rather distorted impression of the geography of the British Isles.

Whatever we may think of the map today, it fulfilled at least one of its original purposes admirably. Government officials were sufficiently impressed to offer Alston the grant that he wanted. Treasury records reveal that the Glasgow Asylum for the Blind was awarded £400 towards the cost of printing Bibles.

POORLY POSITIONED: The dot representing the Dublin, the capital of Ireland, is placed incorrectly, some distance south of its true position.

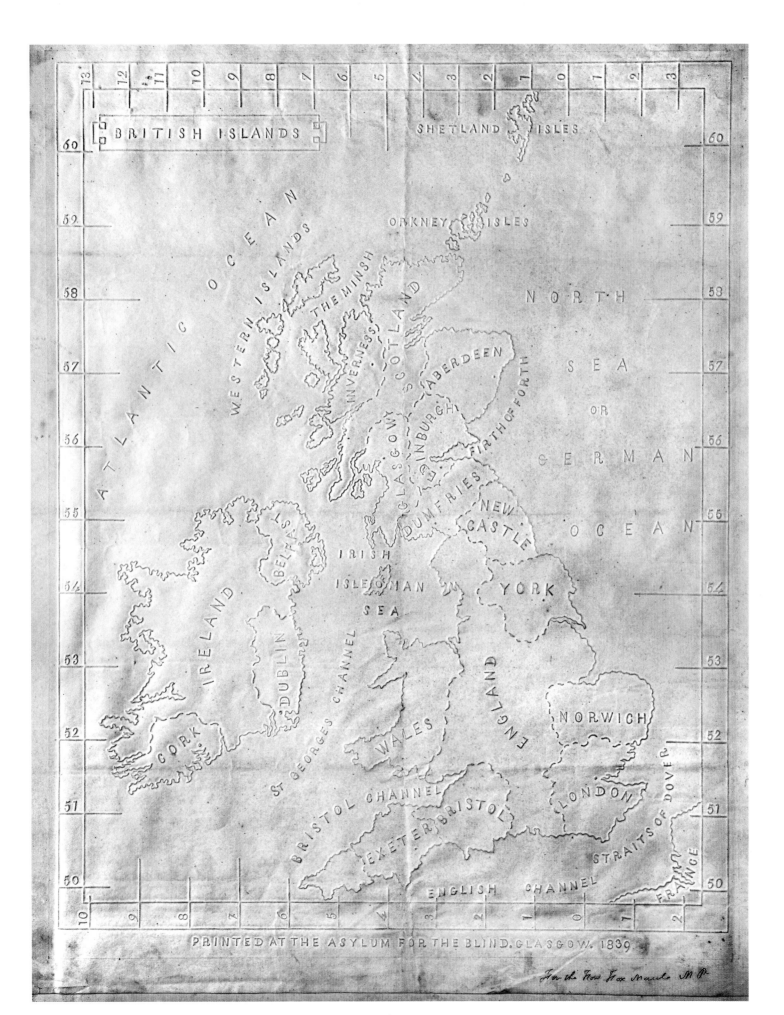

PRINTED AT THE ASYLUM FOR THE BLIND, GLASGOW. 1839

# Perfect for a park
ALBERT PARK, EAST LONDON, 1845

London is a densely populated metropolis, yet it contains many parks and open spaces, large and small. Some of these are very old – Hyde Park and St James's Park were first opened to the public in the seventeenth century – and others much newer. The development of Regent's Park in the 1820s inspired the creation of other public parks in cities and large towns throughout the United Kingdom over the following decades. In London itself these included Battersea Park and Kennington Park in the south and Victoria Park in the east. The latter, named after the young Queen Victoria, was developed in the early 1840s to redress the lack of green space in the districts east of the city centre.

The idea of public parks held a special place in the minds of Victorian social reformers. Recognising that working-class city dwellers had few, if any, opportunities to enjoy fresh air, exercise and respectable entertainment, proponents of urban parks hoped to improve the wellbeing of the 'deserving poor'. As spaces shared by people from different social classes, parks were thought to encourage the less-well-off to learn from the example of their 'betters'. The eastern side of London – traditionally poorer than its western districts – was seen as particularly likely to benefit from new parks.

The designer of this map, J J Maslem, believed that Victoria Park alone could not meet the area's leisure needs. He envisaged the creation of a second park in the districts of Stepney and Bromley-by-Bow, to the south of Victoria Park. This would be called Albert Park after Queen Victoria's husband, the juxtaposition of the two parks being intended as a tribute to the young royal couple and

their happy, respectable family life. Maslem imagined the layout of Albert Park in considerable detail, including proposing new roads to connect it with the City of London and Regent's Park. Determined that no buildings should interrupt the park's interior, he designed rows of new houses (shown in pink on the map) around its edges.

Maslem paid for just twelve copies of this map to be printed. He enclosed two of them with letters to the government's Works Department. In these letters, he explained his vision for the park and urged the government to purchase the land required to create it, which he feared would otherwise soon be swallowed up by housing and industry. Officials did not accept Maslem's suggestion and the vividly-imagined paper landscape of Albert Park was never brought to life.

Two later attempts to create a London park named after Prince Albert, the first between 1850 and 1852 in Islington and the second on Hampstead Heath in 1853, were also unsuccessful. Although none of London's parks bears Albert's name today, Maslem's idea of a fitting companion for Victoria Park has finally come true. The Queen Elizabeth Olympic Park in nearby Stratford – named after Victoria's descendant Elizabeth II and originally developed as a venue for the 2012 Olympic Games – has become London's newest recreational open space.

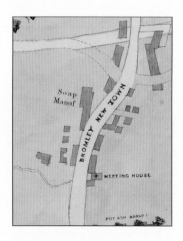

REDEVELOPMENT: Maslem's plan involved the demolition of Bromley New Town, which he considered to be a slum. He proposed to replace its factories and poor-quality housing (depicted here in grey) with new, better buildings overlooking the southern edge of Albert Park.

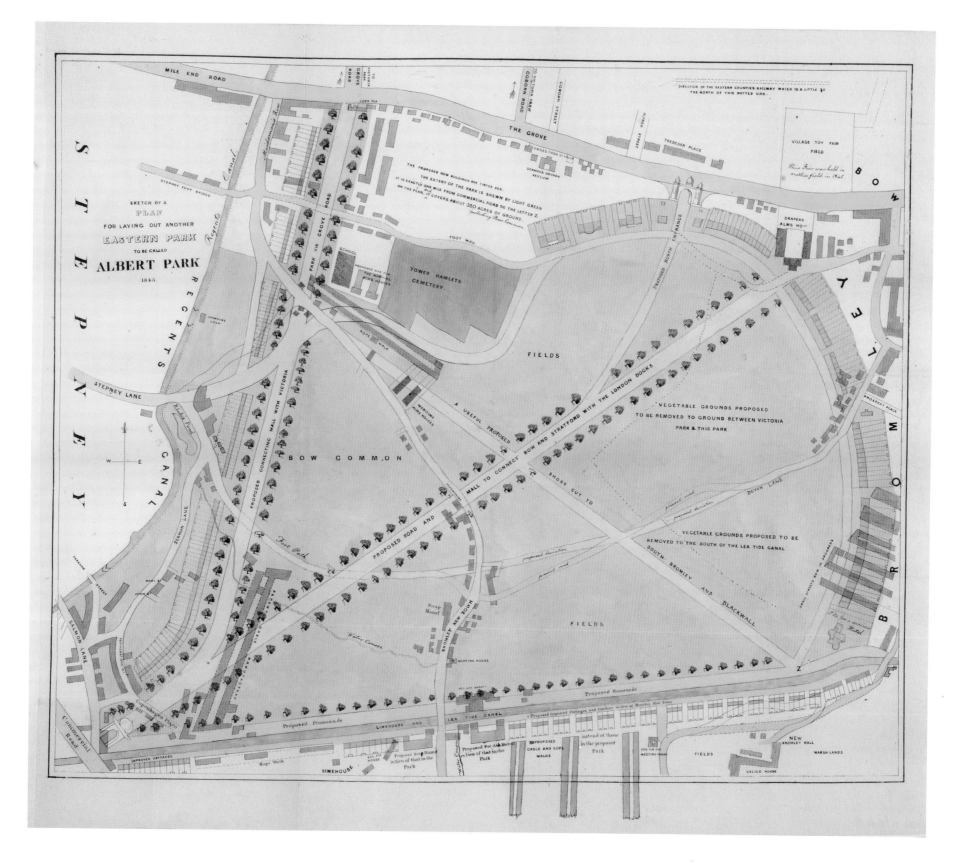

# In the palm of your hand    LONDON, 1851

This map of central London is one of the most unusual records in The National Archives' collections. It is not just a map in the shape of a glove, but a wearable item of clothing, although, as far as we know, nobody has ever worn it.

The map design has been painted onto the leather in ink. Some of London's most famous buildings, including the British Museum, Westminster Abbey and St Paul's Cathedral, are shown as labelled pictures. The largest building, labelled 'Exhibition' is the Crystal Palace in Hyde Park. This spectacular iron and glass building was designed by Joseph Paxton for the Great Exhibition, which took place between May and October 1851. Buckingham Palace is not named but can be identified by the flag flying from its roof. By tradition, the Royal Standard is flown above the palace whenever the monarch (in this case, Queen Victoria), is at home.

Many people are surprised to learn that the historical records of the British government include clothes and other objects. In this case, the glove is a 'representation', or sample, that was sent to the Office of the Registrar of Designs to protect the design owner's legal rights. It now forms part of the historical records of the Board of Trade. Although most design representations are still kept in the original heavy volumes, each containing hundreds of images or samples, this glove is so uniquely valuable that it is housed in a special small box designed by one of our expert conservators.

According to the accompanying register, this design was registered by George Shove of Deptford, Kent, in January 1851. Shove was not a professional cartographer but an artist and decorator. At the time of the Great Exhibition, he owned a business in London's New Oxford Street, but within a few years the business had failed and he was imprisoned for debt. He later became a corn merchant, as his father had been, but died in 1863, aged just 37, leaving a widow and several children.

We believe that Shove designed his map as an aid to tourists, particularly fashionable ladies visiting the capital for the Great Exhibition. Some people also think that it reflects a Victorian equivalent of the stereotype that women find it more difficult to read ordinary maps than men do. There is no evidence that Shove attempted to produce glove-maps commercially and this is thought to be the only surviving example. Although the engraver James Allen issued a printed 'hand guide' supposedly based on Shove's design, that engraving bears little resemblance to the original design other than the general shape of a glove.

Perhaps Shove's vision of maps in the palms of our hands was simply too advanced for the technological possibilities of his time. Fashionable women – and men – visiting London today can use the mapping applications on their mobile phones to find their way through unfamiliar streets.

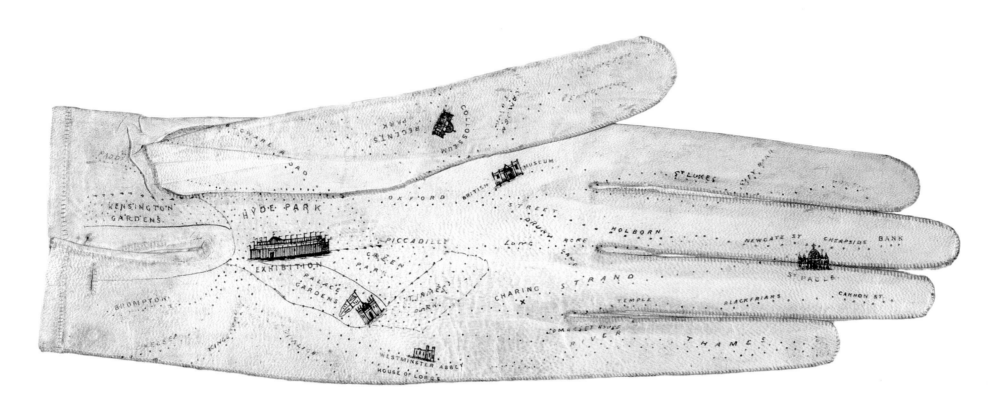

PALACE OF GLASS: The Crystal Palace
which housed the Great Exhibition was
planned and built in just nine months. Built of
cast iron filled in with glass, it was 564m long
and 138m wide. More than six million people
passed through it during the course of the
Exhibition.

MANY-SIDED: The Colosseum, a 16-sided
polygonal building in Regent's Park was built
in 1827 to house a Thomas Hornor's
Panoramic View of London.

LONDON LANDMARK: From 1710 until
1962, St Paul's Cathedral to the east of the
city of London retained the title of London's
tallest building.

# The measure of Melbourne

Like the glove map on the previous page, this square piece of printed fabric is a sample of a registered design. It was created by Allan Arthur Fletcher & Co, a Scottish firm of dyers and calico printers based in Glasgow, and is one of more than 500 cloth designs registered by the company during the 1870s and 1880s. We do not know whether this particular design was ever mass-produced, or even what its purpose was. As it is too small to be a tablecloth or shawl but rather large and coarsely-woven to be a handkerchief, we think that it may have been intended as a bandana or small scarf.

A map of Australia and New Zealand forms the heart of the design but covers a relatively small proportion of the cloth. A series of eight vignettes (alternately depicting ships and landmarks) surround and almost overwhelm the central map. The border, which features a pattern of acorns and oak leaves surrounded by a life-size tape-measure, offers a colourful contrast to the rest of the design. Although the oak leaf, which is traditionally associated with England, seems a surprising choice for a Scottish designer, this type of pattern was in keeping with the taste of the period. Judging by other fabric designs registered in the mid-1870s, flowers, leaves and geometric patterns, often in the deep or bold colours of recently-invented synthetic dyes, were very popular. Although the fabric has perished in a couple of places, the deep, blue-purple colour has faded very little, if at all, over the decades.

In the second half of the nineteenth century, Australia was divided into several separate British colonies. As the map indicates, the bulk of European settlement on the continent was concentrated near the coasts, particularly in the south and east, a pattern that persists today. Melbourne, the capital of Victoria colony, had grown rapidly during a gold rush in the 1850s and 1860s to become Australia's largest and wealthiest city. All four of the landmark vignettes represent buildings or open spaces in Melbourne: the Town Hall, the Post Office, Fitzroy Gardens and the busy thoroughfare of Bourke Street. After the various colonies were united in 1901, Melbourne served as the interim seat of the Australian federal government until 1927 when the planned city of Canberra was ready to take up its role as the permanent capital.

The four ships incorporated into the design remind us both of the importance of maritime trade to the British Empire and of the fact that the Australasian colonies were founded on migration by sea. Hundreds of thousands of British and Irish people settled in Australia and New Zealand during the nineteenth and early twentieth centuries, encouraged and sometimes subsidised by the government. Fletcher & Co may have created this design for the export market, perhaps as a means for the prosperous citizens of Melbourne to express pride in the new land and new lives that they had chosen. It would have been equally suitable for sale in the United Kingdom, prompting those in the mother country to recall relatives or friends on the other side of the world.

CIVIC CENTRE: Two views of Melbourne Town Hall, as seen on this fabric design and as shown in an early 20th-century photograph.

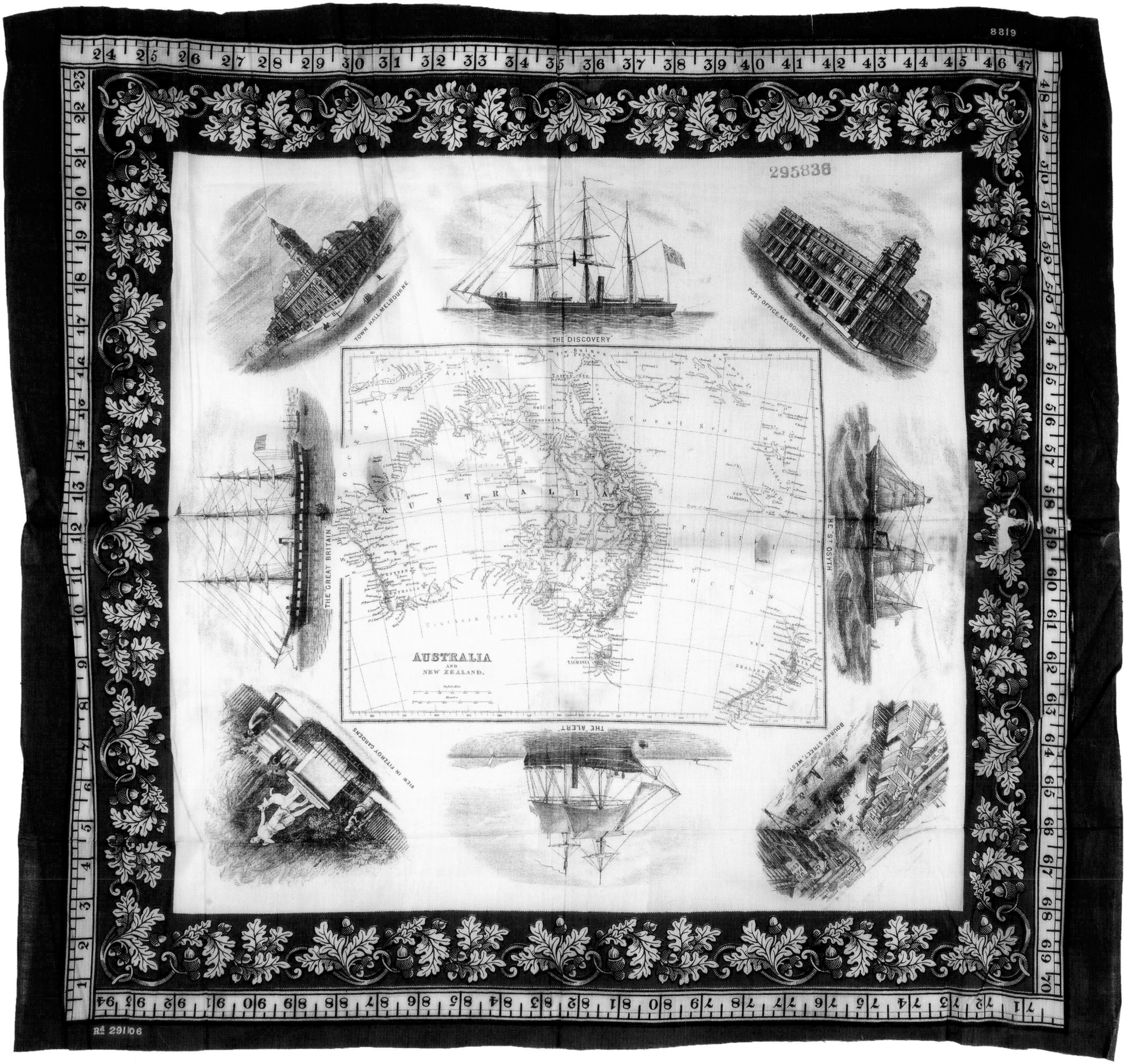
295836
TOWN HALL, MELBOURNE
THE DISCOVERY
POST OFFICE, MELBOURNE
THE GREAT BRITAIN
HE ST. OSYTH
AUSTRALIA
AND
NEW ZEALAND.
AUSTRALIA
TASMANIA
Coral Sea
PACIFIC
OCEAN
Gulf of Carpentaria
NEW CALEDONIA
NEW ZEALAND
THE ALERT
VIEW IN FITZROY GARDENS
BOURKE STREET, WEST
Rd 291106

# An imaginary continent AFRICA, 1886

Sir Henry Hamilton Johnston (1858–1927), usually known as Harry Johnston, was a British diplomat and colonial administrator. During the 1890s, at the height of the 'scramble for Africa' between the European colonial powers, he was a key figure in consolidating British control over some of the United Kingdom's new colonies in southern and eastern Africa. Regarded as a leading British authority on Africa, both during his career and in retirement, Johnston pursued broad scholarly interests and wrote widely about the fauna, flora, history and languages of the continent. Cartography was another interest. He drew many maps while working and travelling in Africa and collaborated with the Royal Geographical Society in London and the Edinburgh firm of John Bartholomew & Sons on the design and production of published maps.

Johnston drew this sketch map of Africa while employed as the British Vice Consul in Old Calabar (now in south-eastern Nigeria). It forms part of a letter that he sent to Sir Percy Anderson, a senior official at the Foreign Office in London, on 13 November 1886. At a time when European control of the continent was still limited to coastal areas and a few portions of the interior, Johnston's letter speculates about the eventual outcome of European plans to colonise the remainder of the continent. Rather than attempting to portray Africa as it actually was, the map illustrates his proposal for how this division into colonies could be established to the United Kingdom's political and economic advantage.

The eventual distribution of Africa between the colonial powers was often quite different from the proposals presented here. This is particularly noticeable in west Africa, where territory was divided between French, British, German and Portuguese colonies (and independent Liberia) in a much more complex way than Johnston had anticipated, and in southern Africa, where a significant portion of the large expanse of territory assigned here to Portugal actually became the British colonies of Northern Rhodesia (now Zambia) and Nyasaland (now Malawi). Ironically, Johnston was instrumental in ensuring that British claims to the latter areas prevailed.

To a 21st century viewer, Johnston's sketch may be reminiscent of the use of maps to illustrate works of fiction. Many authors, especially in the science fiction, fantasy and alternate history genres, have exploited the power of cartography to make their narratives seem more convincing to their readers. Such maps may depict imaginary worlds, fictitious places in the real world, or real places portrayed in a fictitious way. This map is closest to the third category: it depicts Africa not as it once was, or ever will be, but as it might have been had history developed differently. Yet there is a crucial difference between Johnston's contemporaries and the writers of fiction: when authors allow their imaginations free rein, they affect only the pleasure of their readers, but when European politicians and their officials re-drew the map of Africa, their imaginations had a profound and enduring impact on the territories and people that they divided between them.

OUR MAN IN AFRICA: This photograph of Sir Harry Johnston was taken in 1896, when he was British High Commissioner for Central Africa.

54

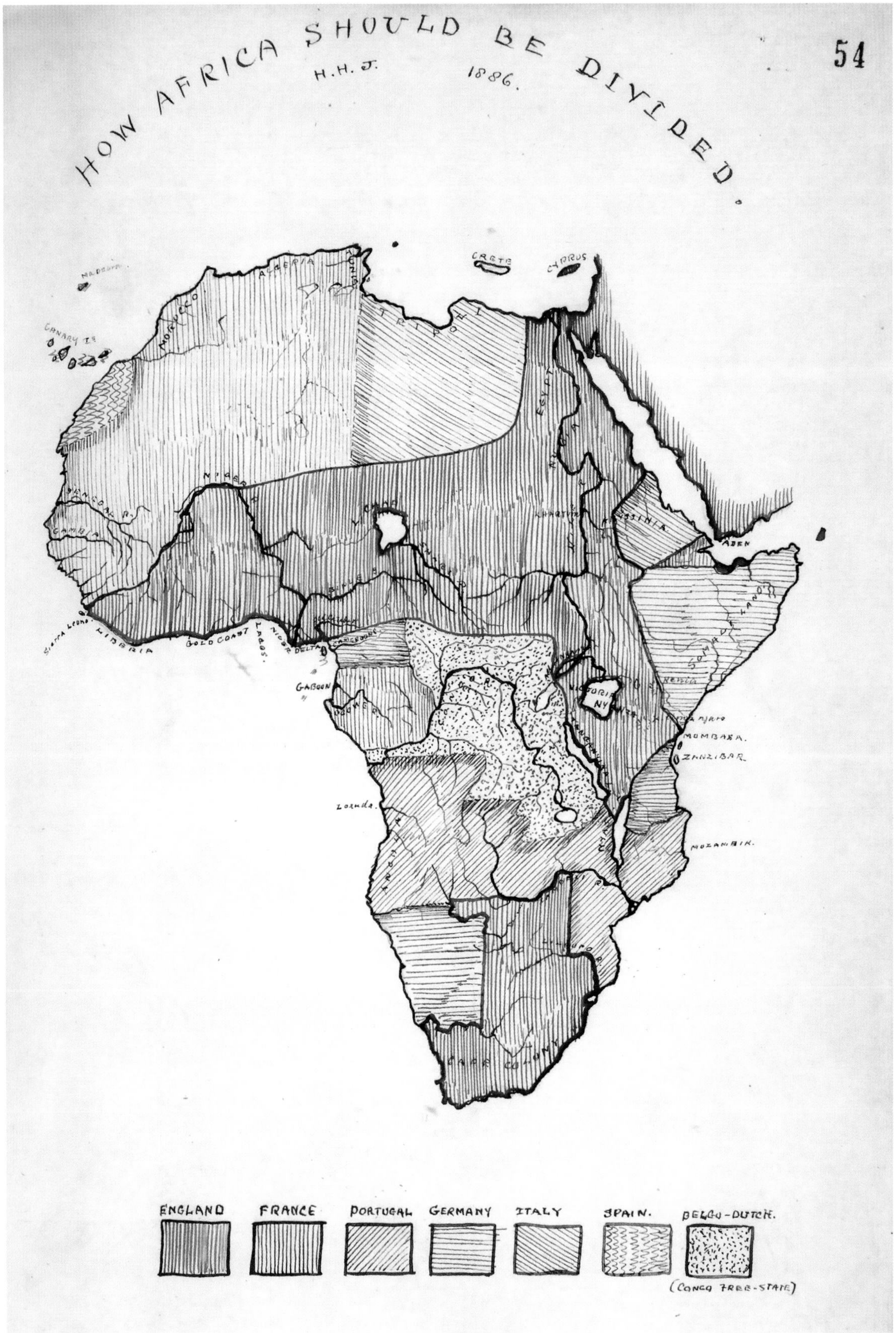
HOW AFRICA SHOULD BE DIVIDED.
H.H.J. 1886.
CRETE
CYPRUS
MADEIRA
CANARY IS
MOROCCO
ALGERIA
TRIPOLI
EGYPT
ABYSSINIA
ADEN
GAMBIA
SIERRA LEONE
LIBERIA
GOLD COAST
LAGOS
NIGER DELTA
GABOON
Loanda
MOMBASA
ZANZIBAR
MOZAMBIK
CAPE COLONY
ENGLAND
FRANCE
PORTUGAL
GERMANY
ITALY
SPAIN.
BELGO-DUTCH.
(CONGO FREE-STATE)

# Cool as a cucumber CENTRAL LONDON RAILWAY, 1911

On 30 July 1900, a new underground railway line opened in London. Running from east to west through the centre of the metropolis, it was appropriately named the Central London Railway, although the original fare of two pence, regardless of the distance travelled, led to its popular nickname the 'Tuppenny Tube'. The line carried 'Underground' branding from 1908 and was taken over by the Underground Electric Railways Company of London in 1913. Absorbed into the London Passenger Transport Board in 1933, it now forms the core of the London Underground Central line, which runs from Ruislip and Ealing in the west to Epping and Hainault in the east.

This advertising poster was produced in 1911. A copy is held at The National Archives because it was registered with the Stationers' Company in London, which was then the usual method of securing copyright ownership. The copyright registration form records that the artwork was produced by John Henry Lloyd, who designed several advertisements for public transport in London around this time.

The poster's bold design cleverly incorporates a simple, diagrammatic route map. In showing the Central London Railway's route as an almost straight, red line, it prefigures today's famous London Underground map, which also sacrifices geographical accuracy for clarity and simplicity. The fact that the Central line is also shown in red on modern Underground maps is probably a coincidence: the Central London Railway was actually coloured blue on many early 20th-century maps.

The summer of 1911 was unusually hot. As July and August temperatures rose to 36°C, road surfaces began to melt and the Times newspaper ran a column listing 'deaths by heat'. By comparing the Central London Railway to a crisp, cold cucumber, the poster implies that a swift underground journey – through some of the deepest railway tunnels in central London – would be far more pleasant than a walk or ride through the sweltering streets on the surface.

Like many advertising maps, this one stretches the truth in several ways. Most obviously, it distorts the railway's route, making it seem straighter and more direct than its true position. A more subtle exaggeration is the 'interchange' shown at British Museum station, which actually involved coming up to the surface and walking to the Great Northern, Piccadilly and Brompton Railway's Holborn station, located some 200 metres away. A genuine connection did not exist until 1933, when British Museum station was closed and replaced with Central line platforms at Holborn. Yet the map's most blatant distortion undoubtedly lies in its promotion of underground trains and tunnels as cool havens from the heat of summer and the busy streets above. It seems unlikely that rush-hour passengers on London's famously crowded Underground network would accept this portrayal today.

WOOD
LANE
SHEPHERDS
BUSH

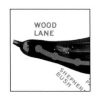

WEST END: Wood Lane station, shown here as the line's western terminus, was opened in 1908 to serve that year's Franco-British Exhibition and Olympic Games. Closed in 1947 and replaced by the more conveniently located White City station, its site now forms part of the Westfield London shopping centre.

Central London Rly.
COME IN HERE!
it's as cool
as a
WOOD
LANE
LIVERPOOL
STREET
SHEPHERDS
BUSH
HOLLAND
PARK
NOTTING
HILL GATE
QUEENS RD
LANCASTER
GATE
MARBLE
ARCH
BOND STREET
OXFORD CIRCUS
TOTTENHAM
COURT ROAD
BRITISH MUSEUM
CHANCERY LANE
POST
OFFICE
BANK
INTERCHANGE STATIONS
THE TUBE
TAKES YOU RIGHT THERE!

LIVERPOOL
STREET
BANK
POST
OFFICE

FAR EAST: In 1911, the eastern terminus of the Central London Railway was at Bank station, named after the Bank of England. The planned extension to Liverpool Street station, shown here as a red-and-white striped line, opened in 1912.

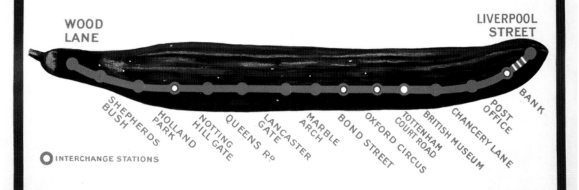

# Race across the Atlantic

Some maps raise more questions than answers and this example is one of them. We understand some aspects of how and why it was made but other details remain mysterious and rather unsettling.

We know that this map was registered for copyright purposes on 2 May 1912 by William Edward Peacock of Islington, north London. Its designer was not Peacock himself but Henry Faulkner of Clapham, south London. The accompanying paperwork and a second design (for a box lid) registered by Peacock at the same time make it clear that this fairly plain, outline map was intended to form part of a board game, and it is likely, but not certain, that it was the board. Since it is only the design of the map that was registered for copyright, we know nothing of the rules of the game or exactly how it was supposed to be played. However, the evidence suggests that it was a racing game between ships crossing the Atlantic Ocean from the United Kingdom to the United States. At this time, Liverpool and Southampton were the main British passenger ports for long-distance voyages.

The inclusion of icebergs on the map (apparently serving as obstacles for the players) and the registration date of May 1912 are clues to what must have inspired the game: the sinking of RMS *Titanic*. The fate of this luxurious passenger ship is well known. She left Southampton on her maiden voyage on 10 April 1912, bound for New York, but shortly before midnight on 14 April, she hit an iceberg, severely damaging her hull, and she sank in the early hours of 15 April. Less than one third of those on board survived. This game design was registered just two-and-a-half weeks later.

The legacy of the *Titanic* has been significant and widespread. On a practical level, the tragedy prompted improvements in maritime safety practices, including lifeboat provision. It has also captured the wider public imagination, inspiring many books, films, as well as conspiracy theories. Set against this context, it is perhaps inevitable that it would attract the attention of a board game designer, too. Nonetheless, a racing game seems a remarkably insensitive treatment of the incident. Unsurprisingly, there is no evidence that the game was ever mass-produced or sold. It seems unlikely that many people would have wanted to play it.

The use of maps and map-like diagrams as boards for playing games has a long history, dating back to the old European 'Game of Goose'. Several modern games, most famously *Risk*, use boards with map designs. Arguably, the various versions of the *Monopoly* board are also maps – albeit rather distorted ones – of Atlantic City, London and other places. The element of purposeful navigation across two-dimensional space invariably gives the two activities of playing a board game and reading a map something in common, even for games where the board bears little or no resemblance to a traditional map.

QUEEN OF THE SEAS: This watercolour painting of RMS *Titanic* was also registered for copyright in 1912.

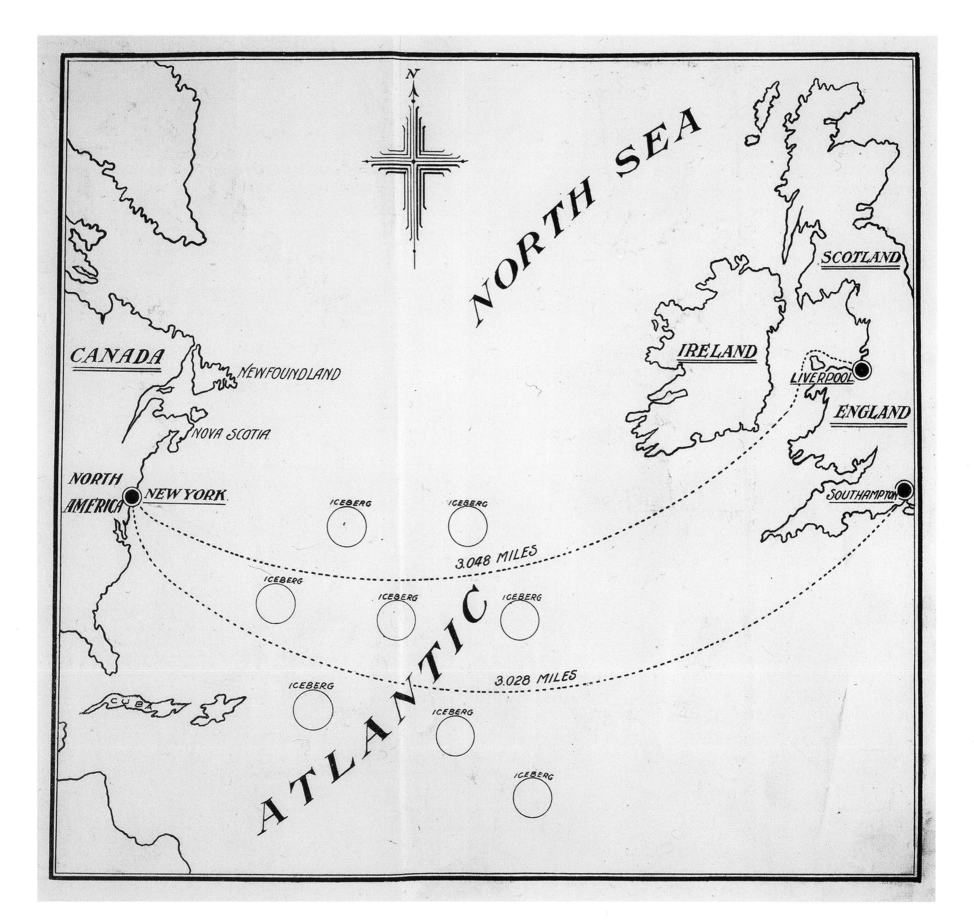

# Make do and mend? EUROPE, c.1939

The National Archives holds almost 2,000 pieces of original artwork created by or for the United Kingdom's Ministry of Information during and immediately after the Second World War. They were created for the Ministry's propaganda campaigns, which aimed to boost morale on the home front and encourage civilians to contribute to the war effort however they could. As well as designs that were actually used in campaigns, such as posters reminding people that 'Careless talk costs lives', the collection includes many paintings and drawings that were never developed further. Some of the designs were commissioned from established artists, including Terence Cuneo, Mervyn Peake and Laura Knight, but others, including this example, are anonymous.

This poster design is based on a map of Europe. The shape of the continent is instantly recognisable. Neither the simplification of the international boundaries – so much so that both Luxembourg and the German exclave of East Prussia have disappeared altogether – nor the fact that many of the boundaries have changed two or three times since this map was made do anything to diminish this.

The map is a collage made from newspaper, card and paint on a base of fairly thin board, the use of different types of paper subtly reinforcing its message that using paper supplies wisely would aid the war effort. As smoking was far more widespread then than it is today, everyone would have recognised that the striking, geometric lettering has been cut out from cigarette packets.

The map is not dated but it is likely to have been made during the first few months of the war. The boundaries are shown in roughly their pre-war positions and the highlighting of the threat posed by the Nazi regime to Germany's neighbours, immediately before and during the early part of the war, would soon have seemed out of date as that threat became a reality. The word 'pacts' probably refers to the Molotov-Ribbentrop Pact, signed between Nazi Germany and the Soviet Union in August 1939, just before the war began.

This design was not selected for use in a campaign. A probable reason for this is that the instruction 'Don't waste paper' is weak compared to such famous wartime slogans as 'Dig for victory' and 'Make do and mend'. Another is that, although the image is powerful and memorable, it is not quite right for the message that it tries to convey. Despite the map being intended for a British audience, Great Britain languishes at the western edge of the map, barely squeezed in and half-hidden by the border. The connection between events taking place abroad and the need to conserve resources at home is not made clearly enough to be persuasive. A design incorporating a map of the United Kingdom or of the British Empire would probably have been a much more effective way of encouraging the British public to use precious resources sparingly.

Carelessness with TYRES – CAN STILL LOSE US THE WAR!

DON'T *speed*
DON'T *corner fast*
DON'T *brake fiercely*
DON'T *accelerate rapidly*
but DO *keep correct air pressures*

ISSUED BY TYRE MANUFACTURERS' CONFERENCE

AN ALTERNATIVE APPROACH: This poster incorporates a stylised map of Great Britain. Although arguably less attractive and interesting than the 'Don't waste paper' map, it is also more conventionally patriotic, and was actually used in a government information campaign.

# Welcome to Smoky Cove

The island nation of Iceland has a long and proud history and culture. First permanently settled in the late ninth century, it boasts both the world's oldest extant parliament, founded in 930, and its most northerly capital city. The city's name, Reykjavík, which roughly translates into English as 'Bay of Smokes' or 'Smoky Cove', alludes to the steam that rises from the area's hot springs. In the early decades of the 20th century, other Europeans regarded Iceland as a sleepy backwater. Although wholly self-governing for its internal affairs, it was part of the Kingdom of Denmark and was largely isolated from international politics. Few British people had even heard of Iceland, except perhaps as an exotic setting for Norse myths.

This situation changed rapidly during the Second World War as Iceland's location became strategically important. After German forces occupied Denmark in April 1940, their British opponents took control of the Danish Atlantic territories of Iceland and the Faroe Islands. The use of Iceland as a naval and air base gave the British, and later the Americans, control of the North Atlantic. Although Iceland officially remained neutral throughout the war – its government refused an invitation to join the war on the Allied side – in practice relations between the Icelandic people and the occupying forces were largely cordial. Thousands of Allied troops remained stationed in Iceland throughout the war and the United States Navy retained a base there for many years afterwards. In June 1944, Icelanders voted for permanent separation from Denmark and became an independent republic. Since then, Iceland has taken a much more prominent role in world affairs, as a tourist destination, financial centre and member of NATO.

This map of the environs of Reykjavík forms part of a small and varied set of maps of Iceland created and collected by the British authorities in 1940 and 1941. Like the board game map on page 241, it is monochrome and (in keeping with its military origins) very plain in appearance. Yet, although it seems very ordinary at first sight, a closer look at some of the text, particularly the curious mixture of road names, reveals its imaginative side. Although a few roads, such as Suðurgata, are labelled with their ordinary Icelandic names, most bear new names bestowed by the British. A number of these are transplanted from famous London thoroughfares, including Charing Cross Road, Bond Street and Piccadilly Circus. Others vary from the practical Red Cross Road to Bo Peep Lane, named after a nursery rhyme character. Some names, such as Grótta Road, are hybrids of Icelandic and English elements. Whether these Anglicised names were intended for secrecy or simply for convenience is unclear. Perhaps the British authorities believed genuine Icelandic names were too difficult for English-speakers to pronounce.

Perhaps the most important – and surprising – example of British military renaming in Iceland is not shown on this map. This was a temporary change to the English name of the country itself. Fearing a potentially-disastrous confusion between Iceland and the similarly-named Ireland in official military communications, the British government decided to refer to the former as 'Iceland (C)' in all of its official wartime documentation.

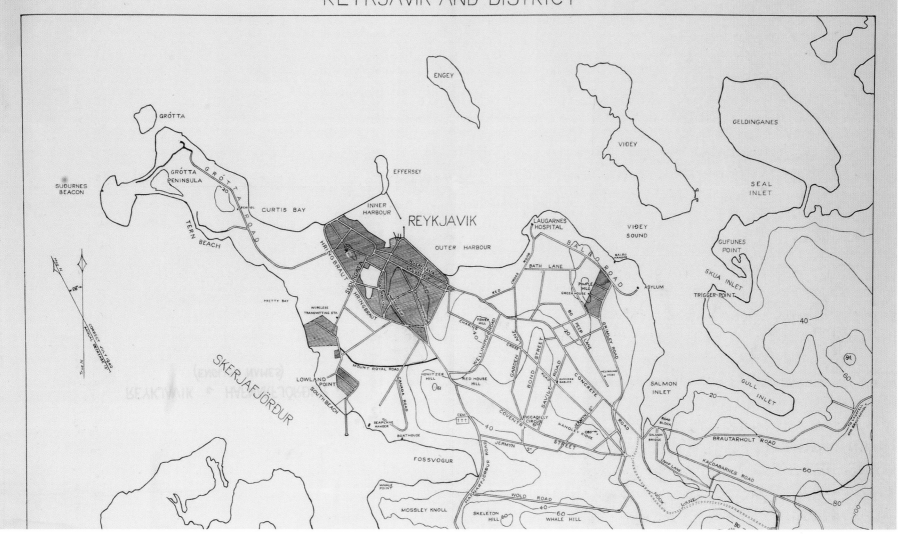

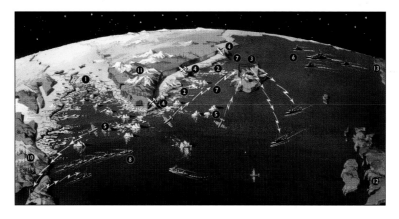

A VALUABLE POSITION: This map showing the strategic location of Iceland in the North Atlantic Ocean was drawn for Neptune, a wartime magazine for British Merchant Navy personnel.

# 'X' marks the spot    

Although the use of maps to record the locations of buried treasure has become a part of popular culture, old treasure maps are much more common in fiction than they ever were in real life. The design of this poster is closely based on a pirate's treasure map featured in the classic children's novel *Treasure Island*, written by Robert Louis Stevenson and first published in instalments in *Young Folks* magazine in 1881 and 1882. Stevenson's story was partly inspired by a map of an imaginary island drawn by his stepson, Lloyd Osbourne.

The poster was designed for a National Savings Committee advertising campaign. The Committee had been founded during the First World War to support the British economy by encouraging members of the public to save money in government-backed financial schemes, such as the Post Office Savings Bank. Its work enjoyed renewed importance during the austerity of the Second World War and its aftermath.

This particular campaign features Captain Bob Saving, a pirate figure whose name and body shape employ a play on words to reinforce the poster's exhortation to save money: a 'bob' was a colloquial term for the shilling, a coin worth one twentieth of a pound or 12 old pence. The shape of Captain Bob's nose is also related to money: it forms a pound sign! The designer has also included a few details on the map that are more relevant to the 1940s than to the 18th-century setting of Stevenson's book. Instead of the traditional pirate hoard of gold or jewels, the treasure chest contains savings certificates. The place names of Poor Man's Peak and Poverty Crag are more subtle allusions to the need for careful financial planning.

Although the cartoon-style drawing places this version of the map firmly in the 20th century, the Treasure Island that it depicts is otherwise identical to Stevenson's in most respects, including the shape and topography of the island and much of the text. Perhaps the map is best considered as an affectionate parody not just of the map in Stevenson's story but of old maps in general. The exaggerated compass indicator, the cartouche in the form of a scroll, and the sea filled with ships and a giant fish are all nods to traditional decorative features. As in all the best pirate stories, the supposed site of the treasure is marked with an 'X'.

The treasure theme of this map makes it an appropriate choice for concluding our book. Not all of the maps in our care are equally visually attractive and not all were made the same degree of skill and care, but we look upon each one as a treasure in its own right. Each map has its own story: it was made for a purpose and played its part in the business of government. Now that they are preserved in The National Archives, each one also has a place in history.

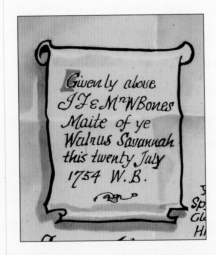

HERE BE TREASURE: The text in the scroll refers to two of *Treasure Island*'s minor characters: Captain J Flint, who originally buried the treasure and drew the map, and William 'Billy' Bones.

# REFERENCES

pp 10-11, **Maps for everyone** – Images: EXT 1/123, part 2 no 17; FO 925/41330; CO 116/77, f 64.

pp 12-13, **Maps for everyone** – Images: WO 78/2761/4; MPA 1/61.

pp 14-15, **Maps for everyone** – Images: IR 128/9/768; MPI 1/230/7.

## EARLY MAPS

pp 16-17, **Early maps** – Images: MPF 1/93; MPF 1/212; MPF 1/68.

pp 18-19, **Early maps** – Images: DL 4/31/9; MPB 1/61/2.

pp 20-21, **View from a medieval monastery** – Main image: E 164/25, f 222r. Insets: E 31/2/1/944, f 32v; E 212/64.

pp 22-23, **Bells, bridges and bog plants** – Main image: MPC 1/56. Related papers: DL 42/12, ff 29-30.

pp 24-25, **A deserted hamlet** – Main image: MPI 1/68/7. Inset: MPI 1/68/8. Related papers: E 314/83, item 16; MPI 1/68/12; PROB 11/22/658.

pp 26-27, **Castles and commons in a Welsh valley** – Main image: MR 1/6. Related map: MPC 1/49.

pp 28-29, **A border line case** – Main image: MPF 1/257. Related papers: SP 50/5, f 175; PC 2/4, f 518.

pp 30-31, **In the shadow of a long-lost palace** – Main image: MPB 1/25/2. Related map: MPB 1/25/1. Related papers: E 315/122, ff 96-97.

pp 32-33, **Beast on the battlements** – Main image: MR 1/14. Related papers: DL 44/75.

pp 34-35, **Sea monsters, galleons and misty mountains** – Main image: MPF 1/68. Related papers: SP 64/1.

pp 36-37, **Summer on the lake** – Main image: MPC 1/33. Related papers: DL 4/13/5.

pp 38-39, **Sea change: a receding coastline** – Main image: MPF 1/212. Related papers: SP 12/254, f 151.

pp 40-41, **Rabbits galore!** – Main image: MPC 1/75. Related papers: DL 1/123.

pp 42-43, **The theatre of the sky and earth** – Main image: MPF 1/127.

pp 44-45, **Around France** – Main image: WO 78/1037/34.

## MAPPING THE METROPOLIS

pp 46-47, **Mapping the metropolis** – Images: MPF 1/287; T 72/13/50, no 3; SP 63/103, no 18; MPK 1/39.

pp 48-49, **Mapping the metropolis** – Images: FO 925/4113; CO 1069/565, no 41.

pp 50-51, **The Knights of St John** – Main image: WO 78/5816, plate 255.

pp 52-53, **Old Father Thames** – Main image: WORK 38/331.

pp 54-55, **Out of proportion** – Main image: CO 700/WestIndies2.

pp 56-57, **At home on the Main** – Main image: MPHH 1/26/4.

pp 58-59, **A ring of bright water** – Main image: MR 1/184/2. Inset: WO 78/419/28.

pp 60-61, **I saw three ships** – Main image: MPI 1/168. Related papers: ADM 1/2012.

pp 62-63, **Town and country** – Main image: MR 1/1200.

pp 64-65, **Fine wine** – Main image: MPG 1/558/1. Related papers: CO 48/149.

pp 66-67, **A city fit for a queen** – Main image: CO 700/SouthAustralia2/1.

pp 68-69, **Eastern capital** – Main image: MPI 1/504/1. Related papers: ADM 101/163.

pp 70-71, **The turtle and the missionaries** – Main image: MR 1/1899. Related papers: FO 228/1615.

pp 72-73, **Separate and unequal** – Main image: MPGG 1/89/1. Related papers: CO 533/142.

pp 74-75, **Bombed out** – Main image: HO 193/2, sheet 56/20 SE. Inset: HO 193/2, tracing from sheet 56/20 SE for 26/27 October 1940.

## CHARTING THE SEAS

**pp 134-135, Charting the seas** – Images: MPF 1/6; C 108/23; MPI 1/95.

**pp 136-137, Charting the seas** – Images: CO 700/South Australia1/1; WO 78/5366; MPI 1/450/3.

**pp 138-139, A medieval mariner's compass** – Main image: MPB 1/38. Related papers: E 122/165/10.

**pp 140-141, Confined to cabin** – Main image: MPF 1/318. Related papers: SP 12/202.

**pp 142-143, The dowry of a queen** – Main image: MPH 1/1/18.

**pp 144-145, Seas of the midnight sun** – Main image: FO 925/4111, plate 11.

**pp 146-147, 'Where ye Fire & Smoake cometh out'** – Main image: MPI 1/26. Insets: ADM 7/688; PROB 4/578. Related papers: ADM 7/688.

**pp 148-149, A shocking shipwreck** – Main image: MPH 1/368.

**pp 150-151, Coffee with sugar: a captured French captain's map** – Main image: MPI 1/118. Insets: HCA 32/176, part 7.

**pp 152-153, Sailing Wolfe to Quebec** – Main image: MFC 1/200. Inset: C 108/23. Related papers: C 108/23.

**pp 154-155, Discoveries of the *Resolution*** – Main image: MPI 1/94. Inset: ADM 55/107, f 137. Related papers: ADM 55/108.

**pp 156-157, After the Mutiny** – Main image: MPI 1/74. Inset: ADM 55/107, f 157v. Related papers: ADM 55/152.

**pp 158-159, Battles of the Nile** – Main image: MPKK 1/64. Related papers: FO 7/290.

**pp 160-161, White on white** – Main image: ADM 352/681. Inset: ADM 7/608, no 88.

**pp 162-163, Gates to St Petersburg** – Main image: MFQ 1/110/5. Inset: MFQ 1/110/9. Related papers: ADM 1/5631.

## NEW WORLDS: EXPLORATION AND THE COLONIES

**pp 164-165, New worlds** – Images: CO 700/NEWYORK13B; MR 1/1771/2; CO 700/BRITISHCOLUMBIA12.

**pp 166-167, New worlds** – Images: CO 700/MassachussettsBay13; FO 925/1011; MPG 1/790/3.

**pp 168-169, Land of great red grapes** – Main image: MPG 1/584. Inset: MPG 1/284. Related papers: CO 1/1.

**pp 170-171, Elephant and castle** – Main image: MPG 1/567. Related papers: CO 268/1.

**pp 172-173, 'Journey to the 5 Indian Nations'** – Main image: CO 700/NewYork13A.

**pp 174-175, 'An Indian a Hunting'** – Main image: CO 700/NorthAmericanColoniesGeneral6/1. Inset: CO 700/NorthAmericanColoniesGeneral6/2.

**pp 176-177, A slave fort** – Main image: MPG 1/224. Related papers: CO 267/11.

**pp 178-179, Rescued by Indians** – Main image: CO 700/Carolina21.

**pp 180-181, Wildlife on the border** – Main image: MPG 1/367. Related papers: CO 5/1098.

**pp 182-183, Kidnap, kauri trees and the underworld** – Main image: MPG 1/532/5. Related papers: CO 201/9.

**pp 184-185, Kangaroo and campfire** – Main image: CO 18/13, f 347.

**pp 186-187, "No white man had ever traversed the country before"** – Main image: MPK 1/422/1. Inset: FO 63/871, pp 240-241. Related papers: FO 63/871.

**pp 188-189, Journeys to the centre of the earth** – Main image: CO 956/537. Related papers: CO 758/60/5.

**pp 190-191, East of Aden** – Main image: MFQ 1/1145. Inset: INF 14/442/7. Related papers: FO 371/69734.

# INDEX

Cuſ

○ Puccantallaha

CUSA RIVER

A HOTT HOUSE

+ Puckana

‡ Wetomkes

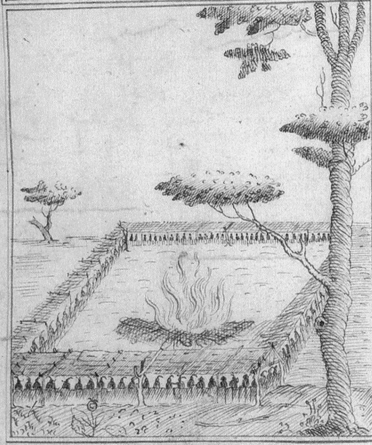

Littetalliſes ○

Fuſato○

W. Gro

Moccoloſus ○ W. Gro
FRENCH
Ockchoy ○ FORT
✝ ✝ ‡ ○ L. Ockcho

Teſtigees... ○

cuſatees ○ ○ Oſwagloes

MOVILLE RIVER

A PUBLICK SQUARE

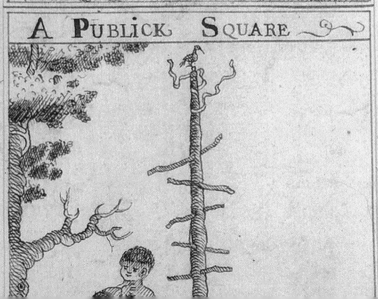

LOWER

Too

Fuctababa c

Atchinalga creek

ATchatalpee creek

Chiſca Taloofa.........○

Wioupkees..or.....○
The Forks..........○

FLIN

Cu